AUDIO REALITY

AUDIO REALITY

MYTHS DEBUNKED
TRUTHS REVEALED

By
Bruce Rozenblit

Published By
Transcendent Sound, Inc.
Kansas City, Missouri

Copyright © 1999 by Bruce Rozenblit

All rights reserved.
No part of this book may be reproduced in any form or by any means without permission in writing from the author.

The information contained in this document is the intellectual property of the author and is for personal use only. Commercial use is strictly prohibited.

Library of Congress Catalog Card Number: 98-91127

ISBN: 0-9669611-0-2

First Printing

While every precaution has been taken in the preparation of this document, the author assumes no responsibility for errors or omissions. Neither is any liability assumed for damages resulting from the use of the information contained herein.

Cover by Visage Design

Table of Contents

What's This Book About? 7
    My Philosopy 9
    What Makes Good Sound? 10

PART ONE: AUDIO SCIENCE
Audio Science 15
    Electricity 16
    Conductivity and Resistivity 21
    Conductors and Connections 22
    Insulators 25
    Capacitance 26
    Inductance 27
    Skin Effect 29
    Reactive Networks 30
    Impedance 33
    The Decibel 34
    Filters 35
    Real versus Reactive Power 35
    Interconnects and Grounding 36
    Balanced Lines 39
    Transmission Lines 41
    Bi-Wiring Speaker Cables 42
    Power Supply Equipment 43
    Power Cords 43
    Isolation Transformers 44
    Damping Systems 45
    Component Compatibilty 48

Voltage Drive Compatibility 48
Speaker Compatibility 49
Vacuum Tubes 50
Feedback and Imaging 52
Servo Systems 54
Acoustics 54
Apparent Loudness 57
Dymanic Range 58
Masking 58
Conclusion to Part One 59

PART TWO: PROJECTS
What Do You Need to Know? 63
    Machining a Chassis 64
    Mounting of Components 65
    Wiring 65
The Transcendent OTL 69
    Checkout 78
Grouded Grid Preamp 83
    Checkout 85
Super Compact 150 Watt Amplifier 89
    Checkout 93
Single-Ended with Slam 97
    Checkout 101
Grounded Grid Cascode Phono Preamp 105
    Checkout 107

My OTL Patent 111

Conclusion 121

What's this Book About?

Audio became a hobby of mine when I was a teenager. Every month, new issues of my favorite magazines were published, which I read from cover to cover. Articles describing the latest developments and products were fascinating to read. Most of all, I loved learning about the technology, the newest circuits, speakers, recording techniques, and hardware. How did they do it? What marvelous innovations would I learn about next month? The articles contained much technical information and taught me many aspects of circuit design.

One consequence of youth is a chronic lack of money. Virtually all of the tantalizing components I read about were financially unattainable. Only one avenue was available that allowed me to have an audio system: I built kits.

Back then many excellent kit options were available. Most were high performance components that were highly regarded in the industry. Sadly, due to changing times and people's interests, the kit powerhouses of the past have disappeared, eliminating the traditional low-cost entry into the hobby. To make matters worse, the prices on contemporary high performance audio gear have risen to extremely high levels. A preamplifier that lists for $3000 is considered in some circles as "moderately priced."

I don't know about your situation, but many audiophiles can't afford to spend $3000 on an *entire system*, let alone one component. This is particularly true for people under the age of 25. The pleasure of enjoying music from quality equipment should not and does not have to be restricted to folks with six-figure incomes.

That's what this book is about. It is a means to get people involved with high performance audio at the lowest cost possible. Audio is a wonderful hobby, pastime, and intellectual pursuit. Much can be learned by building your own equipment. It is something fun to do and, when completed, a high-quality product is yours to enjoy for years to come.

Many myths and misconceptions are commonly held among audiophiles. That seems to be the unfortunate nature of this industry. For those who wish to purchase commercially produced equipment, I will attempt to debunk many falsehoods and clarify misunderstandings with solid, scientific fact, which is the *truth* of my work. All explanations will be simple and won't require any math. I hope to equip the reader with the knowledge required to weed out false claims, appreciate the legitimate, and reap the maximum benefit from your purchasing dollar.

Just as with the kit giants of the past, these projects are not seconds or substandard in any way. They are the actual designs of my company, Transcendent Sound, Inc., that are already in production or will be introduced later. The sonic performance is virtually identical to my commercially produced products. The only difference is that they have been adapted to use off-the-shelf components that are easy for the hobbyist to find, primarily in regard to power supplies. An electronic circuit doesn't care where it gets its power as long as it is of sufficient quality and there is enough of it to go around.

The reader who wishes to purchase assembled equipment is strongly encouraged to study the project chapters. Much information is presented that is nontechnical and concerns the "how and why" of equipment design. Reasons and explanations are given that illustrate my decision-making process, exemplifying why Transcendent Sound products are designed and built the way they are. They represent the *proof* of my work, verifying the many innovations of my designs.

You will notice that none of the projects use any exotic, so-called "audiophile grade" components–for good reason. I don't use them in my commercial products either. All specified parts will give the highest quality performance at the lowest possible cost with no compromises in sound. It doesn't make much sense to spend 500% more to achieve a 1% difference, particularly when you could use that money to buy something else, like speakers or a refrigerator.

Contrary to commonplace reports in today's popular literature, the differences caused by these components are barely discernible, if detectable at all. Yes, some parts do make a difference. Practically speaking, the selection process has a lot more to do with avoiding the "bad" than finding the "best." There is a lot of "good" to choose from at relatively low cost. Once a sufficient level of quality is obtained, any further expenditures produce results that are extremely subjective and subject to interpretation. Often the "difference" detected is not necessarily better or more accurate, just a slight alteration of the perceived sonic experience. For some reason, the magnitude of the perceived benefit is often proportional to the amount of money spent on the device.

I believe much of the hype associated with these and other audio accessories is generated by what I call the "frustrated audio designer syndrome." Those afflicted want desperately to make a contribution to the field by revealing some new discovery that dramatically improves the performance of one thing or another. Unfortunately, they usually only have the ability to swap out a capacitor, change a binding post, or hang some device on the ceiling that claims to redirect cosmic rays. Driven by their overwhelming desire to help, they enthusiastically report of the phenomenal improvements of their great discovery which undoubtedly gets published somewhere. Many of these claims are sincere in intent, but have little or no basis in fact.

A major factor that contributes to this conflict is distinguishing between what one actually hears and what one *thinks* they hear. I can calibrate my voltmeter to give precise readings, but I can't calibrate someone's ears for degrees of sound stage width, voice localization, "dark" tonality and other nebulous characteristics that cannot be quantified in any possible way. How wide is wide and how dark is dark, whatever dark sound is anyway? (I really have trouble understanding when people tell me something sounds bluish or whitish).

The placebo effect must always be considered. It seldom is. Credible psychological and physiological studies are always run with placebo. This is the only way to statistically determine if the results are valid. People are also highly influenced by visual appearance. We generally don't choose our mates based on the sound of their voice but by their appearance. If an audio component looks very attractive, the odds are increased that the listener will think it sounds better.

Successful marketing can implant a preconceived notion in the mind of the listener concerning the quality of performance. The product's reputation and image can convince the listener that the performance is superior before any music is heard. Expectations and visual influences, therefore, play a definite role in the aural evaluation of audio components.

I have had the pleasure of discussing with many audiophiles, both consumers and people in the industry, shared sonic experiences of a similar nature. That is, we listened to a piece of equipment and were able to describe a similar sonic impression. It is possible, therefore, to form a consensus about the nature of how something sounds. How "good" it sounds is another matter. Either you like the way something sounds or you don't.

Example: If you went to the two finest

French restaurants in town and ordered chocolate mousse (assuming you like that stuff) and thoroughly enjoyed each dessert, how can you say which was better? Both differ, but are still wonderful. The best-tasting mousse depends on who is doing the tasting and what kind of mood they are in when the tasting occurs. The only practical answer is that there is no "best" or "ultimate" anything. The problem cannot be solved. So, once I get outstanding results, I go no further and keep my money in my pocket. I assure you that if correctly assembled according to these plans, you can make audio equipment that has equivalent performance to my commercial products (and better than most other commercial products) at a fraction of the cost.

Why am I publicizing my designs for all of the world to see? First of all, anyone that wants to find out what I am doing can buy a unit, take it apart and trace out the circuitry. Hiding the design is impossible. Secondly, most everything I do completely contradicts current audio convention. Since everything I do is thought to be "wrong," obviously no other manufacturer would have any interest in copying my work. Doing so would just validate and endorse my technology. Lastly, let's just say I have something to prove. Becoming a commercial success is primarily dependent on marketing. It doesn't matter what you are selling. You just have to do a really good job of selling it and the profits will come.

I want to be known for my design achievements. Each project presented in this book is markedly different from the status quo of typical high-end amplification equipment. All are genuine alternatives to the standard offerings of the predominant manufacturers. The major product variation you are likely to see between the popular brands is in cosmetics alone, not in design or performance. My work represents truly original creations that produce actual benefits in form and function. The projects presented in this book verify my claims of originality. What better demonstration of innovation could there be than presenting the actual design itself?

I firmly believe that most of the economic woes that high-end audio is currently experiencing are self-inflicted, caused by a severe lack of innovation resulting in no new product for consumers to get excited about. How many times can you repackage something, call it "new and improved," and expect people to part with serious amounts of money to buy it? This problem is greatly worsened by the fact that most of the potential buyers already own a similar product that works perfectly well! Mass marketers have been doing "repackaging" for years with cereal and detergent, but these are consumables, not durable goods. Audio amplifiers usually last longer than breakfast. Many people will only consider replacing an expensive item if it breaks down or wears out and isn't worth repairing, or they can upgrade to something with higher performance and more desirable features. Without innovation any market will stagnate, contract, and eventually could even disappear.

MY PHILOSOPHY

Every designer has an orientation or philosophy that steers his work. Here's mine:

1. Fully 99% of the sonic characteristics reside in the design. Circuit topology controls everything. The designer manipulates inter-stage impedances, time constants, signal headroom, feedback and many other parameters to achieve a desired result. It is not a matter of which is right, but one of achieving a desired result.

2. I totally reject all generally accepted audio convention, both past and present, regarding what should or should not be done. I don't care about doctrines that dictate rules governing use of feedback, class of amplifier operation, variety of tube, types of parts, arrangement of parts, circuit topology, etc. There are no rules. All combinations are possible. I use whatever works best, period. Good results are solely the result of good implementation which emanates from good design. You wouldn't tell a master chef he can't use oregano in a recipe. If he knows what he's doing, a delicious recipe will result. Audio is the same. You just have to know what you're doing.

3. Reliability is everything.

4. Simplicity is the keynote of design. If the parts aren't there, they can't break.

5. All audiophiles come equipped with the most important piece of test equipment there is: a pair of ears. The only person that the component has to please is the person using it. It's *your* system. Once an audiophile called me to talk about buying a used OTL (Output-Transformer-Less) from a previous owner. I asked him to describe the amp and he gave me a very exact description of the construction. Then he kept asking me how it sounded. The amp was sitting in front of him, probably already connected into his system, but he still felt compelled to call and ask me how it sounded. Why couldn't he just turn it on and find out for himself? He had to have a third party advise him. He was looking for a hero to show him the way. My friends, audio gurus make lousy heroes. What I or anyone else says doesn't matter. Your impression is the only one that counts.

6. Audio gear should be designed to share space with human beings. It has to fit in a room and have space left over for people.

7. The less waste heat produced, the better. Few audiophiles live in Antarctica.

8. Mechanical transformer/chassis hum is just as undesirable as hum from the speaker, if not more so.

9. Cost does not directly relate to performance. A superior design is one that achieves outstanding results at the lowest cost.

10. Audio components should have no sound of their own. Instead, they should only reveal the music to the listener to the fullest extent possible. The best amplifiers are those that you don't consciously listen to. You forget that they are there and you are only aware of the music. That's how I feel about it!

WHAT MAKES GOOD SOUND?

For one reason or another, it seems like many people in audio proclaim that they have found THE answer, or the ULTIMATE configuration for the best sound. The purveyors of such thinking are always saying "you can't use this or that" or "you have to use this or that." They are all wrong. This is another one of those questions that cannot be answered. In audio, the whole *is* the sum of the parts. The parts by themselves do nothing. The audio signal that emanates from the output port is the result of the assembly behind it and how all the parts function together as a single unit. That is solely a function of the values of the parts, how they are operated, and how they are connected together. In short, the sound is determined by the design. Accepting this concept changes the question from "What is the ultimate?" to "How do I make the device sound and work in the manner desired?" The task is then one of engineering. That's what I do.

In order to engineer a specific sonic result, it is necessary to understand what amplifiers do and why some sound differently than others. This has been the subject of much debate over the years with no clear-cut resolution. From my years of experience building amplifiers, I believe that I have been able to distill my experiences into a methodology that can be utilized in real world applications.

Music is an incredibly complex form of information. It is dynamic in amplitude and in spectral (frequency) content, both at the same time! The task of the amplifier is to generate electrical power to drive a type of motor (the loudspeaker) to faithfully create an acoustic parallel to the original electrical signal. This is not an easy task when attempting to achieve the highest standards possible.

If a test was run with two similarly powered low distortion amps, one tube and the other solid-state, where a sine wave was run through each at high power, I seriously doubt that anyone could tell the difference. One could then deduce that solid-state and tube amplifiers sound alike. The test fails because it checks for the wrong criteria. Play music through each and a difference can be heard.

Why does the tube amp sound differently from the solid-state amp or for that matter, why does any amp sound differently from another? The answer lies in the world of engineering. I'm sorry to disappoint, but this has nothing to do with magical devices, absolute truths, altered states of reality, or other realms of mysticism. It's about hardware and their operating characteristics called *transfer functions*.

Amplifiers have transfer functions. So do the individual tubes and transistors that comprise them. A transfer function is a mathematical relationship that defines what gets added to or taken away from an electrical signal when it passes through a device. This must be true, otherwise all amplifiers would sound the same. The differences in sound are caused by information being left out and added to the original musical signal.

Individual active devices such as tubes and transistors have their own transfer functions. Specifically, the active device's transfer function defines how the component translates voltage into current. In the case of music reproduction, how the device handles the wildly changing dynamics of the program material also factors in the equation.

Tubes and transistors are made of completely different physical structures and operate on different principals. Therefore, they have different transfer functions. Dissimilar tubes have varied sounds because of their respective transfer functions. Tubes possess other characteristics that cause further changes. When driving large amplitude signals through tubes on the test bench, they can be made to sing. The signal going through the tube causes mechanical forces to act on the internal structures and make them vibrate. The signal can actually be heard projecting out of the tube. This tells me that tubes actually *modulate* themselves. Tubes with rigid structures tend to have more detail. Less rigid tube structures tend to sound softer with less resolution. This reasoning helps to explain why different types of tubes can sound so different even though they may have similar electrical measurements.

A major factor is that a loudspeaker is a nonlinear load for the amplifier. It is a *reactive* load. That means that the current flowing through it and the voltage across it do not follow in lock step with each other. They are actually displaced from each other in time. To make matters worse, the degree of this time displacement is also nonlinear. It changes with respect to frequency and amplitude. Everything is moving around in three directions at once! The amplifier must cope with this changing load and provide a signal to the speaker with all of its harmonic components in the right place at the right time. This is a far cry from driving an 8 ohm resistor on a test bench with a sine wave.

I am absolutely convinced that the most critical power range that primarily determines how an amplifier sounds lies below one watt and quite possibly below 100 milliwatts. I believe that there are tiny subcomponents of the audio signal that are 20 to 40 dB below the program material which distinguish amplifier sound. (See pages 34-35 for a thorough discussion of dB's). How the amplifier handles and resolves these tiny signals at small fractions of a watt, while simultaneously generating the main signals, is where the battle is fought.

Many may not realize it, but a speaker with a sensitivity of 90 dB can play loudly on just one watt. For music to sound twice as loud, the power must increase 10 times. That means our subject speaker will sound twice as loud with 10 watts and twice as loud again with 100 watts. An average sound pressure level of 100 dB is extremely loud. When listening at moderate volumes, the base sound pressure level will probably be 75 to 80 dB. This corresponds to a power level of roughly 50 to 100 milliwatts into our example speaker. That means these critical subcomponents of the sound must lie at power levels of less than 10 milliwatts. This is why I am convinced that the first watt is everything as far as quality is concerned.

High power operation is another matter. In order for quality to be maintained, different characteristics are required from the amplifier. The drive signal must be clean without excessive distortion. This does not mean distortion levels of only tiny fractions of 1 percent are mandated. Actually, 2 or 3 percent distortion will sound fine at high volumes. Things fall apart when distortion rises to over 10 percent. Then the sound gets muddy and breaks up. It is more important that the amplifier generates enough power to achieve the desired sound pressure levels than to try to hold distortion to .05 percent. You can't hear the difference between tiny fractions of distortion when you are trying to pin your ears back with a blast of sound. Often times, the loudspeaker itself will produce distortions of more than 10 percent under these conditions.

Another task that an amp must perform at high power is maintaining resolution of individual voices. When listening to passages of complex orchestral music at high volume, lesser quality amplifiers will tend to merge the individual instruments into one homogeneous sound. The best amplifiers will preserve the voicing of the individual instruments so all of the sections of the orchestra or band can be heard. Tube amplifiers tend to be better at this task than solid state units. I don't know of any test that can be done to measure this characteristic, but it is most definitely audible. I suspect that the manner in which the amp translates voltage into current while driving a highly reactive load has something to do with this.

If I am correct in my assumptions, then steady-state, or static, measurements where a single tone is injected into an amplifier and measured for distortion, reveal almost nothing about how the unit will sound. I am not saying that conventional distortion measurements are unimportant or frivolous. Certainly equipment producing 15% distortion is quite objectionable. What I am saying is that once distortions are reduced to low levels, such as less than 0.5%, the measurements don't really tell us anything. Furthermore, steady state measurements reveal almost nothing about performance with harmonically complex, highly dynamic signals fed into reactive loads, which is a more realistic scenario.

Many alterations are available to the circuit designer to manipulate the way an amplifier sounds. There are techniques that can be applied which will have similar effects in any circuit. The major areas to adjust are input and output impedances, rise time, feedback and damping, and inter-stage impedances.

Amplifiers that operate with no closed loop feedback are sometimes said to have super bass. What is usually happening is that the woofer is operating uncontrollably, and is generating much more bass energy than was originally present in the signal. As feedback is added, the bass tightens up and becomes much more percussive. There is an optimum point where the sound is very realistic and faithful, but it takes some feedback to get there. This is one of those "less is more" situations.

Similarly, feedback affects the sound staging and imaging. With no feedback, images tend to blur, with no precise localization and the sound stage tends to broaden out. Some people really like that effect–I do not. As feedback is added, localization becomes better defined and precise but the sound stage will tend to narrow.

The manner in which an amplifier or preamplifier handles large transients also affects bass performance. Feedback and damping heavily influence circuit behavior under these conditions. When a large bass transient occurs, under-damped circuits can overshoot and actually produce an exaggerated bass sound. My products do not suffer from this problem as they all have a flat response to low frequency dynamic signals. Whenever people hear more bass energy emanating from their speakers, they usually think they are getting better bass because there is more of it. What they often fail to realize is that they might be hearing an exaggerated bass signal. The bass was never there in the first place and was created by the equipment.

Rise time strongly impacts an amplifier's ability to resolve low level, high frequency information. Slowing down the rise time can soften the sound and round off the edges. Speeding it up reveals more detail–sometimes to the point of sounding irritating and harsh. The high frequency damping characteristics integrate with how well the unit handles reactive loads. Some crossover networks are highly reactive and can cause terrible problems for amplifiers.

There are many other things that can be done. The point I am trying to make is that for every design, a series of compromises must be made. Everything you do affects some other parameter in a negative way. There are no absolute rules or answers, just a good set of compromises to achieve quality sound.

It is not possible to make any type of amplifier sound like any other type. Adjustments can only skew performance so much. The range of possibilities with the many configurations available are broad enough that everyone should be able to find something to his or her liking.

PART ONE
AUDIO SCIENCE

*"It's oh so easy to run out of time,
oh so common to run out of patience,
and oh so costly to run out of knowledge."*
—Bruce Rozenblit

AUDIO SCIENCE

When I first started marketing my OTLs, audiophiles would call and ask me to describe the product. I would then proceed to explain the highlights of the circuitry. They would stop me and say, "No, I'm not technical. I just want to know what kinds of parts it has." I soon stopped trying to describe the circuit.

For some reason, many audiophiles have a terrible apprehension of anything that is considered "technical." What motivations could cause these attitudes? I'm sure one is that technical information is considered by many to be totally unreachable by the lay person because it is shrouded in advanced mathematics and mired in incomprehensible abstract concepts. Another possibility is that true high-end audiophiles have broken the shackles and limitations of science and engineering and do not wish to be restricted by such earthly things. This requires using devices and procedures that create perceived benefits but defy all attempts to reveal any measurable differences, even with the finest and most sensitive scientific instruments.

If one is a follower of either of these two belief systems, then one is at the mercy of the mass marketers who have intentions to capture as much of your money as they can. It is to the benefit of all audio consumers to educate themselves so as not to fall prey to the unscrupulous. You don't have to go to college to acquire basic information; just read this book.

Belief systems work like a religion. Followers range from the mildly interested to the fanatic. There is nothing wrong with having belief systems. They serve an important function by providing a means for people to interact and bond together. In the case of audio, getting too involved with a belief system can cause a lot of unnecessary expense. After investing a significant amount of money in equipment, based upon a particular belief system, it becomes very difficult to convince someone that they could have obtained better performance for half the money. The more money spent, the stronger the belief system becomes.

Much of what is taken as proper and correct in this business is nothing more than "he said-she said." The weight given to the information depends on the speaker's image and following. These people often don't provide any justifications for their recommendations or reveal any reasoning to back up their predictions. Neither do shamans. They usually just say, "You should do this because I said so." I am a follower of science and only care what the science reveals. Science teaches us the *mechanisms* of nature. If you understand how and why things work, then you can predict the outcomes of natural events. Instead of relying on someone else's preaching, knowledge can be used to determine what beneficial approach should be taken. Faith is a necessary component of spiritual matters but knowledge will go a long way when selecting hardware.

For some unexplained reason, technical and academic credentials don't seem to be a major factor in labeling someone an "expert" in the often strange world of audio. There are no exams, academic standards, professional requirements, or certifications of any kind required for anyone to design, manufacture, sell, review, publish, or engage in any activity concerning audio equipment. The industry operates like a sixteenth century bazaar. You set up your table under the tent and start selling. Anything goes and most everything does. The types of products sold in this industry range from those with the highest integrity based on solid design and top-notch construction to total fraud. Claims range from the honest and legitimate to statements that are so bizarre that they would never make it into a low budget science fiction movie. I'm sure none of you would ever consider letting the mailman put braces on your child's teeth, or let the plumber remove your appendix, but most don't think twice about buying a $10,000 amplifier from someone who doesn't even have two years of tech school training. Perhaps the mass marketers know something I don't, like how to sell anything to anybody.

The information presented here is scientific fact, all drawn from standard textbooks and professional references. It is not feasible to attempt to present all of the theory that one would receive in a four-year degree program. Instead, I will concentrate on those issues that most directly relate to purchasing decisions commonly made by audiophiles; primarily cables, wire, capacitors, connectors, amplification and power supply equipment. I'm not a speaker guy, so those areas will be omitted. I know enough to know what I don't know. I am not going to say which products are legitimate and which are not. What I am going to do is present fundamental engineering science and then let you decide. No equations and mathematics will be used in the explanations for fear of repelling the people who need this information the most.

The next few sections will concern basic fundamentals of the nature and properties of electricity and electrical signals. The category of science that addresses these subjects is physics. What follows is a highly condensed discussion of the physics of electricity and electrical signals. This reads very slowly. Each sentence can cause one to pause and ponder. It's supposed to. Take your time and don't give up. Scientific material never reads like a novel. You will probably have to read each paragraph two or three times. Depending upon how much science you have been exposed to, the next section probably should not be attempted in one sitting. Odds are that you are not going to fully absorb and comprehend 100% of the concepts presented the first time through. This is normal when studying science. Come back to it in a few days and the obscure parts should become clearer. It is very important that you try to understand this material as it forms a foundation for much of what follows. The information builds on successive sections so it would be best if you didn't skip around. Hold on for a wild ride. Some of you are not going to like what follows.

ELECTRICITY

We use it all of the time. Our lives are totally dependent upon it. Most people don't have the vaguest idea what it is. In order to understand it, we have to examine its three manifestations: the electric charge, the electric field, and electric current. We can then understand where it comes from, what makes it go, and what to do with it.

All matter is composed of atoms. No substance can be broken down into anything smaller than an atom without being changed into a different material. The most principal element of electricity is the *electron*, one of the three major components of atoms. The other two are *protons* and *neutrons*. In an atom, the protons and neutrons form the center called the *nucleus*. Surrounding the nucleus are the *electrons* which move around it in spherical shells, like concentric spheres. Think of placing a soccer ball inside a beach ball at its center and a baseball inside the soccer ball at its center. This arrangement would form three concentric spheres. The areas that contain the electron movement are called *electron shells*. There is a tremendous amount of space between the electrons and nucleus. In fact, atoms are mostly empty space. If an atom was the size of a baseball, the nucleus would measure only 25 millionths of an inch. Studying the behavior of electrons, their fields and motion, will reveal what electricity is.

An electron possesses an *electrical charge*, making it a *charged particle*. The amount of charge is constant and equal among all electrons. There are two kinds of charges, positive and negative, with electrons possessing the negative type. Protons have a positive charge. Associated with the electrical charge is an *electric field*. A field is a type of energy projected into space like a light beam or radio signal. The field emanates from the electron in all directions like an expanding sphere. The farther away, the weaker the field gets. When two or more electrical fields interact, a resultant force is created called the *electrical force* which is one of the fundamental forces of nature. It causes oppositely charged particles to be attracted (drawn toward each other) and like charges to be repelled (pushed away from each other). The electrical force is the glue that binds atoms together to form materials and compounds.

In many metallic substances there are electrons in the outer most shell that are loosely bound to their atoms. The connection is so loose that these electrons freely move at random through the metal, jumping from atom to atom in a totally haphazard fashion. This ease of electron movement is what allows an element to be used as a conductor.

Nothing in the universe moves without the application of energy. Energy from the food you eat allows you to walk, energy from gasoline powers your car, and energy from electricity operates a fan. Because of the ease of electron movement between the atoms of a good conductor, very little energy needs to be applied to coax them to move in any particular direction. The source of energy that powers the aforementioned random movement of electrons in a conductor is *heat*. At ambient conditions of 77 degrees Fahrenheit, a considerable amount of heat energy is present. Only at a temperature called *absolute zero* would this motion stop and that is 459 degrees F. below zero!

By applying energy to a conductor in the form of an electrical field, electrons can be made to flow through it in a specific direction. *An electric current is therefore the flow or migration of*

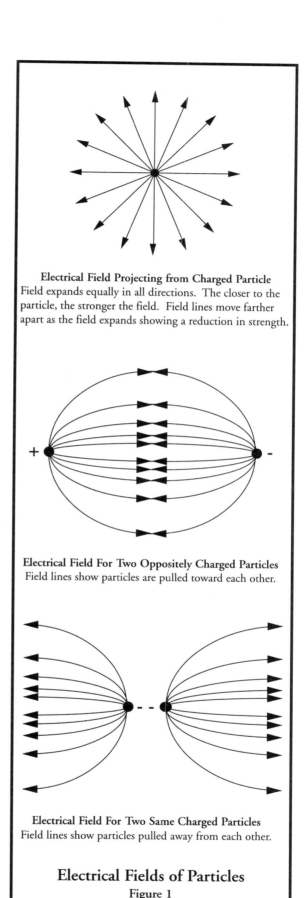

Electrical Field Projecting from Charged Particle
Field expands equally in all directions. The closer to the particle, the stronger the field. Field lines move farther apart as the field expands showing a reduction in strength.

Electrical Field For Two Oppositely Charged Particles
Field lines show particles are pulled toward each other.

Electrical Field For Two Same Charged Particles
Field lines show particles pulled away from each other.

Electrical Fields of Particles
Figure 1

electrons (charged particles) through a conductor. The classical analogy for electric current is to think of a pipe full of water. The water molecules represent the electrons. If there is no pressure on the water, it will not flow through the pipe. Water pressure represents the electric field. Applying pressure pushes on the water and it moves through the pipe. What actually makes the electrons move and how do they move? Answering these questions reveals how to control or modulate electrical current to do useful work.

We must now turn our attention to the static condition of adding excess charges to a conductor where there is no place for them to flow. This is called an *insulated conductor* and the subject is called *Electrostatics*.

Imagine a piece of wire suspended in the air, not in contact with anything. What would happen if we could somehow inject a quantity of electrons into the wire? Where would they go?

The mechanism for causing electron movement in wire is the application of energy. Energy added in the form of heat will only cause increased random motion with no net direction established. The type of energy used to cause electron movement in a particular direction is the electric field. We have already stated that a charged particle has an electric field associated with it. So to generate an electric field, we need to collect a quantity of charged particles.

Static electricity is an example of this collection process. Sometimes, when you walk across a carpet and touch a door knob, you get shocked. The energy from walking causes electrons to be transferred from the carpet to your shoes or vice versa. The direction of transfer is determined by the materials involved. The more you rub your shoes against the carpet, the more charges are transferred and the greater the shock. The charges that are picked up are equally distributed all over the surface of your body. Touch the doorknob with your elbow, forehead, or leg and you will still receive the same shock. The shock is caused when the charged particles leave you and is therefore a *discharge* of electric current from your body. The rate of that discharge is obviously very fast as it lasts only as long as the instantaneous flash of a spark.

Let's suppose we did inject a quantity of electrons into the suspended wire of our example. This would obviously cause a concentration of charges at the injection point. These charges by virtue of their close proximity would generate an electrical field. The resultant field would then act to immediately disburse the extra electrons. They would move away from each other until they were equally dispersed. When they disperse, they jump to the nearest atom and displace an existing electron which jumps to its neighbor causing electrons to move in a chain reaction. In fact, they would disperse until the same number of electrons that were added had relocated to the very surface of the conductor. This happens because *no conductor can maintain a static internal electric field.* Any electrical field inside a conductor will set up forces that will move electrons until the field is reduced to zero. This action is so complete that the extra electrons we added will displace existing electrons until an equivalent number are located on the actual surface of the conductor. The water analogy for this action would be a bucket of water under a flowing spigot. As the bucket overflowed, the water going over the side is not the water immediately dropping into the bucket. The overflow is caused by water being displaced by the additional water added to the bucket.

There is a limit to how many electrons can be added to the conductor. Eventually, the charge becomes so great that the electrons will jump through the air to another object and discharge. The discharge of static electricity when a door knob is touched is an example of this. So is lightning.

Those who have studied physics know that energy cannot be created or destroyed. We added energy to the wire when it was injected with electrons. Where did the energy go? The electrons that moved to the surface of the conductor contain the extra energy. It is held in an electric field which is now projecting out of the surface and is always oriented perpendicular to it. We say that these electrons have a higher *potential* than the electrons inside the conductor. These extra electrons have a *voltage*.

When we say that an electron has a voltage, that is a measurement of how much energy it contains. The higher the voltage, the greater the energy and the capacity to do work. Large residential air conditioners always operate on 240 volts instead of 120 volts. That's because, at 240 volts, the same amount of current can do twice as much work as 120 volts. The delivery of energy is therefore more efficient, allowing more energy to be transmitted in the same size wire. Utility power lines operate at many tens of thousands of volts for the same reason. Returning to our water analogy, voltage can be thought of as the amount of pressure on the water. The greater the pressure, the more force the water has, and the more work it can do. A high pressure stream of water can wash more dirt off a car than a low pressure stream.

The static case just described illustrates the mechanism of charge redistribution inside a conductor. It is this mechanism that causes electrical current to flow.

In order to sustain a current flow through a conductor, a means must be found to maintain a continuous electric field. Some external device must be employed to provide this function. A battery is such a device.

Batteries produce electrical energy through a chemical reaction. They constantly generate free electrons that move to one terminal as they are drawn off by an apparatus or load like a light bulb. Every electron that is drawn off changes the chemistry inside and another is sent to take its place.

The concentration of electrons at the battery terminal is the source of the field. If a conductor is connected between the two terminals of the battery, an electric current will flow. The difference in voltage between the terminals of the battery produces an electric field through the conductor.

An electric field is a manifestation of a voltage. They are, in effect, the same. The voltage across the conductor defines the strength of the field. Field strength is measured in volts per distance. The maintained field is *not* a contradiction with the static case just described. That example refers to conditions where there is no continuous source of electrons and no applied external field. With a maintained field, the conductor acts as a conduit which carries the electric field. The electrons inside of the conductor migrate in an effort to dissipate the maintained field. The field cannot go away because the battery is still connected, so the electrons keep moving. Disconnect the battery and all migration stops.

The direction of electron movement defines *polarity*. Reverse the connections on the battery and the current will flow in the opposite direction. When current flows in one direction only, it is called *direct current* or DC.

Now here is where things get really strange. Electricity moves at practically the speed of light but electric current does not. In fact, it creeps along at a speed of about one inch per second! How can this be? There are two factors that must be examined.

Electric fields are like radio waves in that they both travel at the speed of light. This means that all of the free electrons in the conductor are instantly affected by the field. Remember, a field is like a voltage. So when the field is set up, all the free electrons instantly acquire essentially the same voltage (assuming the wire has zero resistance), and consequently the same energy. What is crucial to understand is that all the electrons are the same. The capacity for them to do work is proportional to their voltage which is now equivalent. Returning to the water analogy, increasing the water pressure at one end of a pipe will essentially cause that same increase in pressure to appear at the other end. The electrons do not have to race through the conductor at the speed of light. It is the electric field, which is moving at the speed of light, that transfers energy through the conductor by raising the voltage of all the free electrons. The electrons barely move in the direction of the field. The speed of the electron current flow is called the *drift velocity*.

Earlier, it was mentioned that, in an uncharged conductor, the electrons freely move from atom to atom in a random fashion. The speed of this motion is billions of times greater than the drift velocity. When a current is established, the electrons still bounce around all over the place, constantly running into other atoms at high speed. An applied electric field causes them to drift in a general direction setting up a current. Think of a storm cloud full of water droplets in the sky. The individual droplets within the cloud may be traveling in various directions at one hundred miles per hour riding on strong internal wind currents. The cloud itself might only be *drifting* along at five miles per hour. The cloud movement represents current flow and the high speed random movement of the water droplets represents electron movement caused by heat energy.

Drift velocity is not fixed. The quoted speed of one inch per second is a typical rate. For any given current flow, the smaller the conductor, the higher the drift velocity. More electrons have to pass through a narrower passageway forcing them to travel at a higher velocity. The larger the conductor, the lower the velocity. In any event, the speed of current flow in a conductor can almost always be outrun by a bicycle.

The case of an *alternating current*, or AC, is then quite the different phenomenon one might envision once electron movement is understood. An alternating current is defined as reversing polarity periodically. Power from the electric utilities in North America has a frequency of 60 cycles per second. The current actually reverses direction at twice that rate at 120 times a second because each cycle consists of equal positive and negative sections. Now this poses an interesting situation. We have already established that electrons migrate very slowly. With AC the electrons in the conductor oscillate back and forth, never really going anywhere because the field reverses polarity before they have had a chance drift very far. A frequency of 60 Hz has a polarity reversal time of .0083 second. If the drift speed is one inch per second, the electrons will drift only .0083 inch before they reverse direction.

The transmission of electric power always requires two conductors. We usually think of them as one going out (the hot lead) and one coming back (the grounded lead). The current flowing in each conductor is equal. The difference between the two is that the current in the grounded lead is at a lower potential, a lower voltage, than the current in the hot lead. The voltage of the grounded conductor is virtually zero. The voltage difference between the two conductors was consumed by the load and signifies all of the energy delivered to it. Therefore, the mechanism of delivering electric power into a load is electrons transferring their contained energy by giving up their voltage into a load. They leave the load with no extra voltage; the load consumed that energy to do work like operate a fan or light a light bulb.

An audio signal is an AC signal. The difference between audio and electric power is that electric power operates over much higher voltages than are typically produced in an audio system. Also, power signals are at one frequency (60 Hz in the U.S.) while audio frequencies vary over the range of human hearing.

An example using audio equipment would be a loudspeaker and a speaker cable. The cable contains a tremendous amount of free electrons, one for each atom. What actually happens is that during the positive half cycle of an audio signal, the amplifier pushes electrons into the cable. Because of the electric field, electrons instantly move from the cable into the speaker. During the negative half cycle, electrons are pulled back into the amp and out of the speaker back into the cable. The process keeps repeating. Because of the huge reservoir of free electrons in the cable, it is highly likely that the electrons coming out of the amplifier never make it to the speaker. Current flow is then a constant exchange of electrons into and

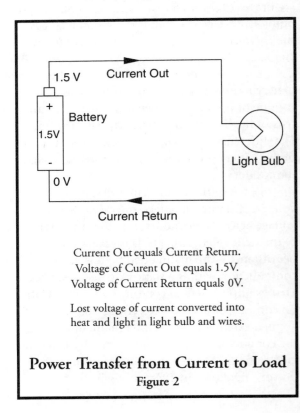

Current Out equals Current Return.
Voltage of Curent Out equals 1.5V.
Voltage of Current Return equals 0V.

Lost voltage of current converted into heat and light in light bulb and wires.

Power Transfer from Current to Load
Figure 2

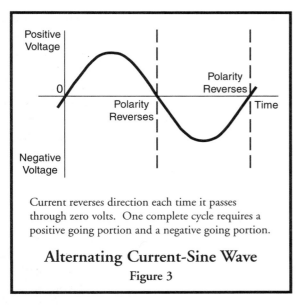

Current reverses direction each time it passes through zero volts. One complete cycle requires a positive going portion and a negative going portion.

Alternating Current-Sine Wave
Figure 3

out of the cable. Voltage is transferred from the electrons to the voice coil, thereby liberating their energy and causing the speaker cone to move. I fear that some cable manufacturer is going to read this and claim that their cables contain special audiophile grade electrons that will improve the performance of any speaker. It wouldn't make any difference. The electrons are merely messengers that transfer the energy of the audio signal, via their voltage, into a load. All of the electrons are the same. The transfer of energy is totally controlled by the electric field. The field is everything in transferring signals.

Electric current flow in a conductor is always controlled by an applied field. The electrons can only move in response to the field. They are totally subordinate to it. Only by changing the field can current flow be manipulated. Thus, the mechanism for altering an audio signal as it travels through a conductor is altering the electric field.

Audiophiles have often told me that their cables have to be connected in a certain direction because they were "broken in" that way. Unfortunately, this is an incorrect assumption. Since audio is an AC signal, the direction of the current flow changes with each half cycle. Half the time it is traveling from left to right and half the time it is traveling from right to left. Therefore, there is no net direction to the current flow. Current flow is equal in both directions. Consequently, reversing the cable connections has no possible effect. By the way, stepping on cables or bending them doesn't do anything to the sound either.

CONDUCTIVITY AND RESISTIVITY

Conductivity and resistivity are inverse properties. A good conductor is a poor resistor and a good resistor is a poor conductor. What causes resistance? Why do some materials make better conductors than others?

The ability of a material to act as a conductor is set by its atomic structure. We can't change that. It's a natural property. The best conductor is silver. Copper is second best, but it is very close to silver. Actually, copper is only 6% less conductive than silver. That's not much of a difference. Increasing the cross sectional area of a copper wire by 6% makes it equivalent to silver wire as far as the electrons are concerned. We have already determined that the electric field controls the behavior of the signal, not the electrons.

Any current flowing through a conductor will cause it to heat up. As the electrons travel, they constantly smash into other atoms. From classical physics, the energy of a moving body is called *kinetic energy*. When the electrons slam into an atom, they transfer their kinetic energy to the atom and make it vibrate. The vibrational motion of atoms causes heat. The more they vibrate, the hotter they are. This is why a small wire gets hot when a large current passes through it. The current density goes up and the drift velocity increases, both causing many more collisions and significant heating.

The hotter a conductor becomes, the greater its resistivity. That's because there is increased random movement in the free electrons as they are excited by the additional heat. The more they move around, the harder it is for them to flow in a given direction as they keep slamming into more atoms. Cooling a conductor decreases resistivity by reducing the random electron movement. When temperatures approach absolute zero, some materials actually loose all resistivity and become super conductors. I hope I didn't just create a monster so that now some cable manufacturer is going to come out with a refrigerated speaker cable. Hold on to your wallets!

Because the electrons keep bumping into atoms, they cannot accelerate to higher velocities as they travel. The collisions keep electron drift velocity low. Therefore, drift speed is so low because there are just too many road blocks in the way.

The primary factors that determine how much resistance a conductor has are its resistivity, its cross sectional area, and its length. These relationships are all directly proportional. Increase the cross sectional area by two and the resistance drops by a factor of two. Double the length and the resistance increases by a factor of two. Use a material with twice the resistivity and the resistance doubles. Materials that make good resistors have an atomic structure in which the free electrons are slightly bound to their atoms. They have a lattice structure that presents lots of collisions to the movement of electrons. *The quantity of resistance,* measured in *ohms,* is that which opposes the flow of current in a conductor. For any given resistance, a specific voltage is required to push a specific amount of current through it. Double the resistance and twice as much voltage will be needed to push the same amount of current through the conductor. That is Ohms Law.

The heating of a conductor or resistor represents a transformation of energy. The heat energy that is liberated is taken from the electrical energy of the current. The debit of electrical energy is accompanied by a corresponding drop in voltage. We say that a resistor has a voltage drop across it when a current flows through it. The amount of energy given off as heat is exactly equal to the amount of energy lost in the voltage drop. The flow of energy is always conserved.

CONDUCTORS AND CONNECTIONS

The technology behind this area is very mature. Switches, relays, terminal blocks, cords, and all manner of connection devices have been in use for more than 100 years with every conceivable type of signal imaginable. Furthermore, electrical connections are made in every type of environment from under water to outer space and at frigid temperatures to blast furnaces. A tremendous variety of materials, platings, alloys, and fabrication techniques are available for the designer to utilize.

Coupling an audio signal in the benign environment of a residence is fairly easy to do compared to the rigors of interplanetary space travel. However, the mass marketers make it seem as if this is the most difficult task in the world requiring exotic materials (read expensive) and top-secret proprietary technology. Before we pass any judgements, let's investigate how the non-audio world makes electrical connections.

The most widely used material for electrical conductors is copper. The type of copper used is called electrolytic copper, ASTM type B5. ASTM stands for American Society of Testing Materials. They set the manufacturing specifications for just about everything. B5 copper is the benchmark for conductivity as all other materials are gauged against it. It is 99.95% copper, .04% oxygen and .01% miscellaneous impurities. It is usually *cold drawn* when formed into wire. That means it is pulled through successively smaller dies until it reaches the desired diameter. No heat is applied during the drawing process. Cold drawing gives it more tensile strength.

There is another type of copper wire that is much more expensive. It is called Oxygen Free High Conductivity, or OFHC. The ASTM designation is type B170. It has a fairly impressive name indeed. The content is 99.95% copper without the oxygen and .05% other impurities. Its primary application is that it better tolerates severe cold forming operations when being fabricated into intricate parts. Its name would tend to make one believe that it would be a great material for audio applications. However, the conductivity of OHFC is *exactly the same* as standard type B5 copper. You just can't judge a book by its cover! It pays to always read the specs. If OFHC was 100% pure copper, the maximum possible improvement in conductivity would only be .05%. This is such a minuscule improvement I doubt you could even sell it to the Department of Defense. I'm sure the increase in price is far greater than five cents on the dollar. In any case, the electric field is the same and therefore the audio signal is unaltered.

Copper tarnishes when exposed to the air and forms a black oxide. Once the oxide sets up, it doesn't keep getting progressively worse, but maintains a steady thickness. If exposed to moisture and carbon dioxide (outdoor applications), it acquires the familiar green patina that you no doubt have seen on buildings and statuary. These oxides actually protect the copper underneath, but they make electrical connections difficult and soldering practically impossible. Therefore, for electronic applications, copper wire and connectors are always plated with another material. Before we discuss plating methods, we need to investigate just exactly what an electrical connection is.

If you examine a highly polished piece of metal under an electron microscope, the surface would look like a jagged mountain range. When a connection is made between two such pieces, the peaks on the surfaces would make random contact. Most of the contact area is actually open space. If the connection is under sufficient pressure, the contact points are forced into each other and a bond is formed called a *cold weld*. The contact points really do weld together. You probably have experienced this many times yourself. Have you ever tried to loosen a metal screw from an unpainted metal object? This does not have to be from an electrical connection. The screw might be small but you have to use all your strength to make it move. It suddenly breaks free and makes a pronounced "snap" noise when it first moves. Then it unscrews very easily the rest of the way. The usual reaction is "What kind of a gorilla tightened this screw down?" The snap you heard was the welds breaking. The screw was not over tightened. Over time, maybe years, the constant pressure caused the welds to become more numerous and stronger because metal tends to flow under those conditions. Sometimes the welding is so pronounced, you have to drill the screw out in order to get the thing apart.

This type of connection is called a *gas tight* connection because atmospheric gasses cannot penetrate the contact areas and cause them to oxidize and deteriorate. The screw may tarnish and look bad, but the actual connection is quite good. High pressure is required for this type of connection to be effective.

The traditional way to implement a gas tight connection is with a *binding head screw*. This type of screw has a wide head with a rim that projects under the outer circumference. The rim of projected metal has the ability to cut through any surface oxidation on the terminal or wire placed beneath it, thereby ensuring an excellent connection. It is extremely important to achieve a high quality, highly conductive connection when working with power wiring. A poor connection will get hot, degrade and corrode, causing it to become hotter still and even cause a fire. Terminal blocks with binding head screws have been in use for many decades but have all but disappeared from the high-end audio scene. By far, the most common material used for this and most other types of electrical connections is nickel plated brass.

Brass is a very old and useful alloy. It consists of copper with a 35 to 40 percent tin content. Its conductivity is only about one fourth that of copper but that is easily compensated for because connectors have very short lengths and the cross sectional area can be easily increased as needed. The big advantage it offers over copper is that it is much harder and has much greater tensile strength. Copper is too soft to be used as a connector in many applications.

One of the problems with brass is that is tarnishes very easily, hence the need for plating. Nickel is the most commonly used material. Why?

Nickel bonds to brass very well when electroplated. Nickel also is extremely resistant to oxidation. You don't have to worry about cleaning it. It is hard and durable, preventing it from wearing off during repeated fastenings. The conductivity is only about one fourth that of copper, but the plating thickness is so thin, that its contribution to resistance is negligible. A few extra micro ohms aren't going to make any difference.

Gold is a highly popular plating material in audio. Its conductivity is very high and it is virtually immune to all forms of oxidation.

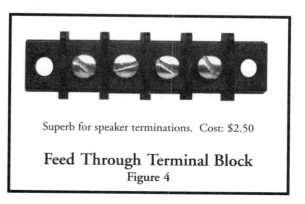

Superb for speaker terminations. Cost: $2.50

Feed Through Terminal Block
Figure 4

Gold is also very soft and cannot withstand much abrasion as it will wear easily.

There are different classes of connections that require different technologies. Two major categories are high pressure/high power and low pressure/low power connections. High pressure/high power connections are primarily found in electrical switchgear where hundreds of amps of current are switched from 480 volts up to tens of thousands of volts. The closest analogy to that type of connection is the speaker binding post albeit the power levels are many thousands of times lower. These types of contacts require great mechanical strength to withstand repeated operations. The materials used for the contacts may be subject to surface oxidation. They are selected for their ruggedness and longevity in this application. The surface oxidation is not a problem because when the switch closes, an arc is formed which literally burns it off allowing for a good metal-to-metal connection. In very high voltage applications, the arc is so powerful it actually vaporizes contact material and the switches are designed to allow most of it to condense back on the contact, thereby recycling it.

The case of the binding post is similar only because the relatively high pressure and rotational action when tightened down cuts through any surface oxides on the terminal facilitating a good metal-to-metal connection. The question then arises: does it matter what material the binding post is plated with since mechanical action alone will create a good connection? The plating material need only provide a durable bearing surface and control oxidation to the extant that it doesn't interfere with the connection.

Many audiophiles feel that a plastic (usually nylon) insulated binding post is cheap and ineffective. The internal construction of the so-called cheap post is usually a solid one quarter inch brass shaft set in a nylon collar. The plastic around the metal is functional. It provides insulation on the order of 1500 volts. The hexagonal part that is tightened has a heavy threaded brass bushing inside which makes the connection with the terminal. The brass parts are plated with either nickel or gold. The consumer should ask: Why is this system any different from any other? Once a solid connection is made at the terminal, what possible benefits can be obtained by having an all-metal top piece?

Another consideration regarding the connections made with binding posts is what type of terminal is used. I use standard tin-plated copper terminal lugs that I buy at the hardware store for about 25 cents each. That choice shouldn't surprise you by now. For a lot more money, "audiophile grade" lugs are available. They sometimes cause more problems than they solve. First of all, they can be so thick that they often won't fit inside the binding post. Their thickness makes them so stiff that it becomes difficult to tighten the post down without using a wrench. Compounding this problem is the fact that many speaker cables are so rigid and inflexible that they place a large twisting force on the connection, mandating the use of a wrench to apply enough torque to keep the binding post from unscrewing.

When cables with reasonable flexibility are used with a conventional sized lug, no wrench should be necessary. Finger tightening should suffice. If additional torque is still required, a nut driver should be used, as an open-ended wrench can easily apply too much torque and twist off the binding post.

The case of the RCA jack is completely different from the binding post. It definitely falls into the low pressure/low power category. The current levels are in the low micro amp range and there is no screw type device to

Heavy-duty binding post, rated at 30 amps, 1500 volts.
Cost: Gold plated-$5.00, Nickel plated-$3.50

Insulated Binding Post
Figure 5

increase contact pressure. These conditions indicate that an oxide-free surface would be desirable. The low contact pressure minimizes wear. Gold is therefore a suitable, if not ideal, candidate for a plating material on an RCA jack. This does not mean that a nickel-plated jack will sound any worse. I personally can't tell any difference. The use of gold increases the reliability of the connection particularly in corrosive environments and it certainly doesn't hurt anything. Nickel is often used as a sub-plating for gold. It facilitates good adhesion between the gold and brass. Some believe that a nickel plating is damaging to the audio signal, but more times than not, whenever you see gold there is nickel underneath. The RCA jacks that I use in my equipment are gold over nickel over brass.

Selector switches are another type of low power/low pressure contact. By far, the most widely used material is the silver-plated brass contact. The brass provides a strong base and silver gives it the lowest contact resistance possible. Silver is also soft, so low contact pressure is necessary to keep it from wearing out.

The problem of silver oxidation must be addressed. It is handled by the mechanical nature of the switch. One half of the contact is shaped like a blunt knife, the other like a flat band. When the switch is rotated, the flat band is dragged across the knife edge and literally cuts through any oxide. This is called a wiping action. A reliable, low-resistance connection is then assured. I have examined many rotary switches on old equipment and, upon looking at the contacts, it is plain to see a dark tarnished contact with a shiny, clean narrow stripe through it. The action of rotation cleans the contacts with each pass. It is good maintenance to exercise a rotary switch at least two or three times a year to prevent too much oxide from building up. Leave an amp in a damp basement for 15 years, and the switches will require more aggressive cleaning with chemical agents.

Gold is sometimes used as the plating for switch contacts and so are gold/silver alloys. The ultimate choice is dependent upon the application–primarily with regard to the environment in which the switch will be used. I use switches with silver-plated brass contacts in my equipment and they work just fine.

INSULATORS

In an insulator, like rubber or plastic, there are no loosely bound electrons hopping from atom-to-atom at will. All the electrons are tightly bound to their host atoms. It would require a tremendous amount of energy to get them to move. So much so, that the insulating material would be destroyed in the process. Materials that make good insulators generally make poor conductors of heat because of the lack of free electron movement.

An insulator must do three things. It has to be long-lasting in the environment that it is used, it has to have sufficient mechanical strength to withstand any abrasion and pounding it might experience, and it must provide enough dielectric strength to prevent a short circuit at the voltages used. That's all any insulation does, no more, no less. Insulation itself has virtually no effect on the propagation of electrons through a conductor or the nature of the electric field that is moving the electrons. It only serves as an impenetrable boundary to the electrons to keep them from going where they should not. Different types of insulation can contribute to minute differences in capacitive effects but these are negligible at audio frequencies.

Stationary Blunt Edge Contact / Movable Flat Band Contact

Movable contacts are spring loaded and press against blunt edge contacts. Rotating switch drags movable contacts across blunt edge contacts thereby removing oxide build up and making a good connection.
Cost: $1.60, enclosed type-$3.00

Rotary Switch Contacts
Figure 6

CAPACITANCE

There are three and only three constituents that comprise all passive electric networks. We have already discussed one and that is resistance. The other two are reactive properties called capacitance and inductance. Capacitance relates to electric fields and inductance to magnetic fields. An electric field can store energy, like a battery. Capacitance is the property that describes the storage of energy in an electric field.

A capacitor has a simple structure. Imagine two parallel metal plates one inch square. Aluminum foil would work nicely. Separate the plates and sandwich a one inch square piece of plastic wrap between them. Press the assembly together so there is no air in between the plates. You just made yourself a capacitor! The plates are commonly referred to as *electrodes* and the plastic wrap is called the *dielectric*. When a voltage is impressed across the capacitor, an electric field is set up between the plates. This field stores energy. The amount of energy storage is determined by two things. One is the capacitance which is primarily controlled by the dimensions of the structure and the dielectric material. The other is the voltage across it. So the two requirements for capacitance to be present are two electrodes separated by a dielectric. The units of measurement for capacitance are the microfarad and picofarad (one millionth and one trillionth of a farad, respectively).

In a commercial capacitor, the internal structure is composed of a film of a conductor and a dielectric, similar to an aluminized plastic tape. These films are highly specialized products and are only manufactured by a few firms in the world. There are many types of films made out of a broad range of materials. Browse through any electronic parts catalog to get a feel for the variety. The capacitor companies buy the films and wind them into the body of the capacitor. Then they attach leads and encapsulate the unit in a durable housing.

All capacitors have a voltage rating. Whatever unit that is selected must be rated to withstand the maximum voltage that will be impressed across it. Otherwise, failures can occur.

Do coupling capacitors affect the sound? Absolutely yes! I can attest to a definite notice-

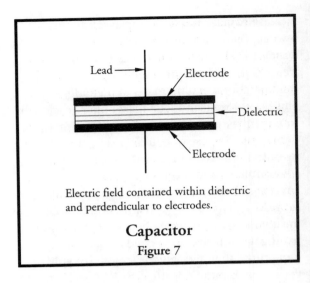

Electric field contained within dielectric and perdendicular to electrodes.

Capacitor
Figure 7

able improvement when changing out a lessor performing unit with a better one. The principle change I notice is that the sound is smoother and less brittle. I do not detect any spacial or dimensional changes in the sonic image. I can confidently say that there is an improvement in the tonality. I can also confidently say that you should be able to find good-sounding capacitors from between $1.50 and $3.00 each. Experience has shown me that film type capacitors generally sound better than those with metalized film construction. I have no idea why one type sounds better than another and have no information to offer.

Different dielectric materials change the capacitance of any capacitive structure. The mechanism for this is that the molecules on the very surface of the dielectric become polarized when immersed in an electric field. That means one of their electrons moves slightly to cause one end of the molecules to be more negative than the other. The degree of electron movement is less than the size of an atom. The term for this effect is called a *dipole*. The polarization of the dielectric sets up an electric field that opposes the field across the capacitor. The net effect is to reduce the total electric field across the capacitor. The more the dielectric reduces the electric field, the greater the capacitance of the capacitor.

Air is a dielectric. Two bare wires aligned parallel to each other with a space in between constitute a capacitor, albeit a really small one. The farther apart the electrodes, the lower the capacitance for any given arrangement.

All wiring possesses capacitance. There is *stray* capacitance between wires and the chassis they are enclosed in. There is also stray capacitance between individual wires. These quantities are all very small and are usually negligible at audio frequencies. The crucial determinates in setting these capacitances are the length of the interconnecting wires and their proximity to each other. The type of insulation has virtually no effect on altering internal wiring capacitance because the distances between the wires predominate. The use of a printed circuit board can greatly reduce these effects because it provides for very short, direct connections between components. Long cable runs between circuit stages are the biggest contributor to stray capacitance. In critical applications, the physical layout of the circuit must be arranged to minimize stray wiring capacitance.

All capacitors block DC. When an input is designated as "AC coupled," that means a capacitor is in series with it. Capacitors therefore only pass AC. The resistance a capacitor has to the flow of AC is dependent on the frequency of the signal. The higher the frequency, the lower the resistance. This relationship is directly proportional. A capacitor is 1000 times less resistive at 20,000 Hz than it is at 20 Hz. Large value capacitors are called *electrolytic* capacitors and their internal structure requires that they are polarity sensitive to DC signals. Install one backwards and they act like a short circuit and will self destruct.

INDUCTANCE

Whenever current travels through a wire, a magnetic field is produced. A time-varying current (AC) produces a time-varying magnetic field. The magnetic field emanating from a straight conductor takes the form of concentric cylinders aligned with the conductor's longitudinal axis. Whenever a time-varying magnetic field passes through another conductor it *induces* a current that flows in the *opposite* direction. If the conductor is formed into loops, each loop induces a reverse current in the loops near it. Since the induced currents flow in the opposite direction of the original current, the conductor becomes more resistive. The effect is like trying to paddle a canoe upstream against a river current. The stronger the current, the harder you have to paddle to move forward. The conductor therefore becomes more *inductive* and the assembly is said to have more *inductance*. The more turns, the greater the inductive effect.

A major difference between the action of a magnetic field and an electric field is that the magnetic field has a directional component associated with it. Two nearby conductors can be physically arranged in such a way as to cancel any mutual magnetic effects. This can be accomplished by twisting wires together or rotating their axis 90 degrees in one plane.

If a magnetic material, like a piece of iron, is placed inside a looped conductor, the magnetic field is made many times stronger. This characteristic is called *permeability*. Air has a permeability of one. Some magnetic materials have a permeability that ranges into the tens of thousands. The inductive effect is proportional to the permeability. Strengthening the magnetic field increases the ability of each loop to induce reverse currents in its surrounding loops. The inclusion of a highly permeable core in any inductor dramatically increases the efficiency of the assembly, requiring far fewer turns of wire to achieve the same inductance.

Inductors for loudspeaker crossovers can be purchased with air cores or magnetic cores. The air core inductors are much more expensive because they require many more turns which necessitates using large wire to keep the resistance down. The advantage of using an air core inductor is that it is totally linear with high currents. A magnetic core inductor can saturate

Inductor constructed of four layers of enameled copper wire wound over a cylindrical magnetic core.

Inductor
Figure 8

and become nonlinear with high currents. This can be avoided by designing a core that is big enough to handle the worst case current loads without going into saturation.

Just as in the case of the electric field, the magnetic field stores energy. If you have ever unplugged a motorized electric appliance while it was in operation, you probably saw a spark jump from the outlet to the plug. In order for a spark to jump through the air, a high voltage must be present. When you pulled the plug, the magnetic field in the windings of the motor collapsed very quickly. The speed of the field's movement as it contracted induced a voltage spike which propelled the spark. The visible discharge is then the liberation of the energy stored in the motor's magnetic field.

Induction is the mechanism that makes transformers work. In a simple transformer, there are two separate coils called the primary and secondary. They are wound on a core of magnetic material. The two coils are electrically isolated from each other. When an alternating current flows through the primary, a corresponding alternating magnetic field moves through the magnetic core and transfers most of the energy of the primary current into the secondary. The replication of the current from primary to secondary can be extremely accurate with the proper configuration. Transformers are used to change voltages, provide isolation from the power line, and change impedances.

Sufficient DC injected into a transformer winding will cause the core to saturate and become nonlinear. That's why it is important that bias levels in output tubes be closely balanced to prevent premature saturation in the output transformer as it usually doesn't take much DC to overload the core. The single-ended amplifier has a special type of core that is specifically designed to withstand a heavy DC flow. Output transformers originally designed for push-pull operation cannot be successfully used in a single-ended design.

The magnetic field inside a transformer is almost wholly contained within the core. There is always some leakage. A certain variety called the toroid has virtually no leakage. Transformers are also susceptible to external magnetic fields. Critical low-level applications require special magnetic shielding to minimize corruption from external fields. Output transformers can pick up hum from nearby power transformers. This is a common problem.

The only possible way to alter the output signal of a transformer with an external device is to introduce a time-varying magnetic field into the core, or project it across the coils. That requires another inductive device (something with coils of wire) and a power source to provide the needed energy. In addition, due to the directional component of magnetic fields, any interaction is highly dependent on the physical orientation of the external field source. Passive devices that contain no power source cannot possibly affect the output signal as they have no means to generate a magnetic field. It would be possible to construct a device that had an internal coil that could, through mutual inductance, draw some minute amount of power from field leakage off the core. Distance, orientation, available field leakage and turns ratios cannot be controlled. There is, therefore, no way to implement this scheme successfully even if some beneficial sonic effect could be derived from the extracted power, which it cannot.

The unit of measurement for inductance is the *Henry*, usually expressed as millihenries. The resistive nature of inductance is the inverse of capacitance. The higher the frequency, the higher the resistance. This relationship is also directly proportional with frequency. Inductors pass DC and consequently are subject to power

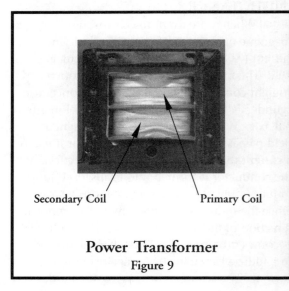

Power Transformer
Figure 9

dissipative effects. The coils of wire inside them contain resistance which will convert some voltage into heat. Also, the magnetic core material can only hold so much energy before it saturates. These two conditions make it necessary to rate inductors for a maximum current level they can safely handle without risk of damage. Also, they are rated for a maximum circuit voltage because the insulation system between the coils and core is designed to withstand only so much voltage.

Magnetic fields have one other property that must be addressed. They exert a force on any conductor that cuts through the field. This is the force that makes motors turn and transformers hum. It also is the force that moves loudspeakers in and out. The study of these forces almost exclusively applies to electric machine theory and does not mandate further discussion. The area of concern that I would now like to address is inductive effects in cables.

SKIN EFFECT

One of the most widely used terms in cable talk is *skin effect*. What really is skin effect?

Any single conductor experiences a degree of self inductance. Alternating currents generate magnetic fields that create reverse currents that vary through the interior of the conductor. This causes the effective resistance to vary through the cross-sectional area of the conductor with respect to AC signals. The increase in resistance is caused by the distribution of current flow which is reduced at the center. The resistance at the circumference is the same as with DC and is not impaired. Hence, the net effective resistance of the conductor is increased. This effect does not apply to DC. The next questions that must be answered are: how much of a change occurs and what factors modify the change in resistance?

The two primary factors that control skin effect are the diameter of the conductor and the frequency of the signal. The smaller the diameter of the wire and the lower the frequency, the lower the skin effect. The type of conductor is also a factor. The higher the resistivity, the lower the skin effect. Aluminum has less skin effect than copper. Silver has more. A high resistance conductor like nichrome wire hardly has any. Let's look at some practical examples to get a feel for the magnitude of this.

A typical size for a hookup wire is 22 gauge. It has a diameter of .0253 inch. At 20 kHz, the effective resistance due to skin effect increases by 0.5%. That isn't much of a difference. For example, a ten-inch long wire looks like a 10.05-inch wire to a 20-kHz signal. The increase in resistance is therefore insignificant.

A common size for speaker wires is 12 gauge which has a diameter of .081 inch. Skin effect is much more pronounced with the larger wire as the effective resistance at 20 kHz is 33% greater. The initial calculation however, doesn't reveal the whole story. A typical length of speaker cable is about ten feet. That means there is 20 feet of conductor because two are required. The DC resistance of 20 feet of 12 gauge wire is .033 ohm, which is an extremely low amount. The action of skin effect increases that to .044 ohm at 20 kHz, an increase of only .011 ohm. Again, this is an insignificant increase in resistance.

So when does skin effect become dominate enough to cause problems? It is a highly important parameter to consider in the design of electric utility high voltage transmission lines because the cables can be as big around as your wrist and run for miles. These cables may carry hundreds if not thousands of amps of current. If they are too resistive, a lot of power will be wasted. Another case is in the design of high amperage switchgear busses that often carry upwards of 1000 amps. They are made of huge bars of copper or aluminum which can cause equipment failures if they overheat when skin effect isn't taken into consideration. These two examples involve signals of only 60 Hz where the size of the conductor is what makes skin effect a potential problem.

At very high frequencies, such as those used in radio transmission, skin effect becomes a dominant factor. The problem is doubly compounded in a high-power radio transmitter such as one used for a commercial FM radio or television station because now the conductor has to handle high current and high frequency. These applications necessitate special dedicated use cables to bring about an efficient transfer of power through them. Microwave transmitters are even more difficult and require conductors, called wave guides, that have special geometries to efficiently transmit power.

The examples illustrated so far have nothing in common with the types of signals generated in a stereo system. It could be reasoned that skin effect is not a significant parameter when transmitting audio signals through cables. The frequencies are too low and the conductors are too small to cause any material problems.

Many cable manufacturers claim that their cables use tiny wires grouped together which greatly minimizes skin effect since the individual strands of conductors are so small that the skin effect goes away. The small conductors are bare. There is no insulation separating them. Consequently, they are all in intimate contact with each other. How can they function as individual conductors if they are all connected together? I submit that they cannot. Furthermore, the data for calculating skin effect have no entry or compensation for stranded cables. Large power cables are always constructed of many multiple strands to achieve flexibility so they can be wound onto a spool and installed in the field. The criteria in determining skin effect is the overall diameter of the cable assembly, not the size of the individual strands.

The speaker cable I use is made up of 260 individual strands of tiny copper filaments. It clearly sounds better than 16 gauge, 20 cent-a-foot lamp cord. (By the way, I paid $1.50 a foot for that cable at a retail store and it was not on sale.) I am unable to identify any specific mechanism associated with using finely stranded cable that makes it sound better than lamp cord, but I only spent $35 so I'm a happy customer. My best guess is it has something to do with the self-inductive properties of the cable.

Here is an interesting experiment that can be used to investigate whether or not skin effect in speaker cables is a factor in audible performance. Ten cables of size 22 gauge are roughly equivalent to one 12 gauge cable. You can buy a 100-foot spool of ten conductor, 22 gauge cable for about $40 (Alpha type 1180C). By using one length of cable for the hot lead and one for the ground, two speaker cables with a length of up to 25 feet can be made. Simply strip the ends of the individual wires and connect them all together. This configuration will dramatically reduce skin effect. The cost is only 80 cents per foot and that can be reduced if you share the cable with a friend. There are dozens of multi-conductor cables with a wide variety of configurations that the audiophile can purchase from parts distributors. Most of these can be bought in 100-foot spools for $30 to $150. Many speaker cables cost over $100 per foot.

Another interesting speaker cable project would be using tined copper braid for the conductors. It is available in either tubular or flat configurations and can be purchased for less than $2 per foot, depending on size. The braid will have to be encased in heat-shrink tubing to provide insulation. Use the very flexible kind, 3M type VFP-876. This will add a little under one dollar per foot. An inexpensive electric hot-air paint stripper can be used to shrink down the tubing.

Considering the extreme range of geometries and configurations available, what does make dissimilar speaker cables sound differently? There must be something else going on that alters the electric field. There certainly is, and the answer involves studying the subject of *reactive networks*.

REACTIVE NETWORKS

Capacitance and inductance are electrical properties that store energy in the form of electrical and magnetic fields respectively. The natural process of storing and releasing this energy takes time. There is always a rate of energy delivery into and out of any reactive component. The charging rate of the reactive component is controlled by the resistance of the charge source. The higher the source resistance, the longer the charging time. Using a water analogy, it takes longer to fill a bucket of water if the spigot is partially closed. The partially closed spigot creates additional resistance that opposes the flow of water. The reduced water flow takes more time to fill the bucket. A full bucket of water represents a fully charged reactive device.

Because a finite amount of time must pass to charge and discharge these devices, they have the capability to alter the voltage of the signal. They can actually cause the signal to be displaced in time. The mechanism for the time displacement is the rate of charging the device.

When a reactive device charges, the voltage across it develops as the charging process continues. For instance, a capacitor cannot instantaneously jump from zero volts to 10 volts just as you can't instantly fill a bucket of water. A certain amount of time must pass for the bucket to fill. Depending on how fast the water flows, the fill rate could be a couple of minutes or a couple of seconds. A slow fill rate is indicative of a slow charging process caused by a high-resistance charge source. When using an AC signal, if the charging resistance is sufficiently high, the voltage across the capacitor cannot change fast enough to stay in lock step with the AC signal. This causes the voltage across the capacitor to be delayed from the AC signal. The voltage across the capacitor has to catch up with the voltage from the source. It constantly lags behind. The signal across the capacitor has been displaced in time.

When the rate is too slow to keep up with the changing AC signal, a frequency roll off occurs. Example: Charge a capacitor with a 10 volt square wave through a sufficiently large resistance such that it took longer to reach 10 volts than the duration of the square wave. The maximum voltage attained across the capacitor would never arrive at 10 volts. The square wave would then return to zero at the mid-point of its cycle. With the square wave at zero volts, the capacitor would discharge and the cycle would then repeat.

This is an example of a low pass filter and it causes a time shift of the signal. The filtering effect is evidenced by the rounding of the corners of the square wave. The frequencies affected by the time shift are primarily in the range of the roll off. The maximum time shift caused by any one individual reactive component is one quar-ter cycle of the signal. Whenever an electronic circuit experiences a frequency roll off, signals in the attenuated range are displaced in time. The steeper the roll off, the greater the time displacement. Any time displacements within the audio band will alter the harmonic structure of the musical signal, which is highly undesirable.

Audio cables always contain resistance, capacitance, and inductance. These are the three properties that constitute an electric network. They have the capacity to create time shifts and frequency roll offs of the audio signal. This, I believe, is the major factor that contributes to the varying sonic performance of different cable assemblies.

When inductance and capacitance are both present in a network, the possibility of oscillation exists. If stimulated with the proper signal, a network that causes an oscillation actually generates spurious new signals that ride on top of the original signal. This is called *ringing* and must be avoided as it can cause serious colorations. The frequency of the oscillations are totally dependent upon the values of inductance and capacitance in the network. Some cables are highly inductive and should be investigated for this possibility. If any oscillations do occur, they probably happen in the ultrasonic or even radio frequency region. The characteristics of the amplifier also must be considered because it may or may not be able to dampen out these oscillations.

The larger the wires and the closer they are

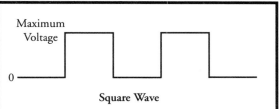

Square Wave

A square wave is an AC signal that stays at 0 volts for a period of time and then almost instantaneously jumps to a maximum voltage. It stays at the higher voltage for a period of time and then instantaneously drops back to 0 volts. The cycle then repeats. Square waves contain many high frequency harmonics and are useful for testing the bandwidth and phase accuracy of amplifiers.

Square Wave Routed Through Low Pass Filter

Square wave is highly misshaped. During charge portion of signal, waveform is curved indicating time required for reactive components to reach maximum voltage. Discharge portion is the inverse of the charge portion. The curved waveform indicates a loss of high frequencies–a low pass filter.

Filtering Effects on Square Wave
Figure 10

together, the greater the capacitance. This follows the structure of a capacitor because the larger wires have more surface area and the closeness reduces the thickness of the dielectric. The smaller the wires and the farther apart they are, the greater the inductance. A common 300 ohm twin lead television antennae cable is an example of this type of structure. It consists of two small diameter wires widely spaced by a plastic dielectric.

No matter what the cable designer does (assuming any design took place), there will always be reactive characteristics to the cable assembly. Procedures that reduce capacitance increase inductance. Procedures that reduce inductance increase capacitance. Neither can be completely eliminated.

Any cable can be modeled or simulated by an assortment of resistive and reactive elements (passive devices) called an *equivalent network*. This model alone cannot predict how the cable will affect any particular sound system. The network functions as part of a system which, for a speaker cable, comprises the device driving the cable (the amplifier), the cable, and the device loaded to the cable (the speaker). Similarly, an interconnect cable reacts with a line level source (a preamp), and a load (the amplifier). A loudspeaker can also be modeled by passive devices into an equivalent network. The affect any cable has directly relates to the interaction of the cable's equivalent network with the signal source and load.

Obviously, cables that possess the minimum of inductive and capacitive reactance should impart a minimum of coloration to the sound. Likewise, the length of the cable should be proportional to the changes since reactance is calculated by units of cable length. Perhaps a good benchmark test for a speaker cable would be to position an amplifier directly behind a speaker and connect the two with a few inches of wire. This has to be the configuration of minimum interposing reactance.

Understanding these characteristics provides a scientific justification for why cable XYZ creates a certain sonic change in one system while that change does not materialize in another. The mechanism for affecting these sonic changes is influenced by how the reactive properties alter the electric field. The reactive properties are primarily determined by the physical configuration of the conductors and insulators, not the materials used. Claims that attribute sonic changes to the subatomic behavior of electron flow are highly suspect and probably false.

The science of quantum mechanics describes the behavior of subatomic particles. The first thing that the student must realize about the field of quantum mechanics is that it is so difficult and abstract that no one understands it. The mathematics required are so advanced that a graduate degree in math is needed just to acquire the tools to investigate it. As difficult as it is, there is one lesson that we can pick up from this field. Quantum mechanics teaches us that it is not possible to determine how any one subatomic particle will behave while under the influence of an external field. It does tell us the *probability* of behavior of a group of subatomic particles. That means we can determine with a high degree of accuracy what

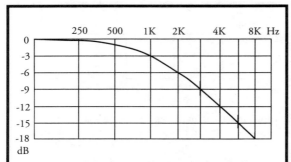

Plot of 1kHz, 6 dB per Octave Low Pass Filter

Graph show attenuation as frequency increases. Vertical axis depicts signal strength in dB's. Negative sign shows signal reduction relative to low frequencies where there is no attenuation. Horizontal axis depicts increasing frequency. Phase shift of signal becomes significant at frequencies greater than 1 kHz and reaches a maximum of one quarter cycle or 90 degrees at 4 kHz.

A 12 dB per octave filter would be at -12 dB at 2 kHz and -24 dB at 4 kHz. The slope of attenuation is therefore twice as steep. The ultimate phase shift would reach 180 degrees at 4 kHz.

Low Pass Filter Characteristics
Figure 11

percentage of available free electrons will behave in a certain manner while under the influence of an electric field. We cannot, however, determine precisely which ones will exhibit the behavior. Claims that boast of specific electron manipulation caused by subatomic structures cannot possibly be verified by any means possible and are suspect, to say the least. In any event, the electric field provides the stimulus to which the electrons respond and the electric field is acted upon by the reactive properties of the cable. Because of the thousands of equipment and cable combinations, there is no reliable way to predict how any particular arrangement will ultimately perform. You don't know what you're going to get until after you get it.

I would like to add that, when I speak of sonic changes, I am referring to changes in tonality and the character of the sound. Qualities such as bright or harsh or muffled can usually be agreed upon by several listeners and often times measured with instruments. When you start saying the first chair violinist moved three feet to right and the sound stage of the percussion section contracted by two feet, there is no way two people will ever agree with such statements. It is pointless to argue something that cannot be verified by any means possible. This is like trying to explain your personal experience of seeing green to someone else. The perceived change may be real or it may be the result of gastric disturbances or a reaction to prescription medication. Like I said before, you either like the way something sounds or you don't.

Interconnect cables also can make changes in the way a system sounds. The varying parameters that control the degree of change are the output impedance of the signal source and the input impedance of the receiving device. Output impedances can vary from one ohm to several thousand ohms. Likewise, input impedances can vary from 10K to over 100K ohms. These factors greatly influence how any interconnect will sound in any particular system.

A somewhat disturbing reality of this discussion is that cables which affect the sound, either by design or accident, are not neutral. They are active participants in shaping the response characteristics of the system. I have to ask the question, "Are cables supposed to do this?" As a designer, I try to make my equipment as neutral as possible, within the confines of any particular configuration, in order to achieve maximum clarity and faithful reproduction. A cable analogy would be a conduit for transmitting a signal from point A to point B without disturbing it or changing it in any way. I suspect my philosophy is not shared by many in the industry.

IMPEDANCE

The quantity of resistance applies to the opposition of current flow. When only DC signals are present, resistance is the only factor that effects current flow once steady state conditions have been met and all the reactive components have charged. When AC signals are present, the quantity of *impedance* is used to describe how devices oppose the flow of current. AC signals are time varying and cause reactive components to constantly alter their state of charge. Actually, resistance is a subset of impedance as all passive devices possess an impedance of which resistance is always a component. Therefore, impedance contains properties that oppose the flow of current *in addition to resistance*.

In the previous section, capacitance and inductance were introduced. They both have the characteristic that their opposition to current flow changes with frequency. More correctly stated, the impedance of capacitance and inductance changes with frequency. The quantity of pure resistance does not change with frequency.

One might ask, "Why make the differentiation between resistance and impedance?" What must be understood is that the mechanism that causes the impedance of reactive components to change always produces a phase shift between the signal current and voltage. The magnitude of the phase shift is determined by the impedance of the circuit providing the signal that is charging the reactive component. The impedance of the reactive device alone does not cause the phase shift. Any phase shift is always the result of the combination of the impedance of the signal source working into the impedance of the reactive device. Impedance can then be used as a tool to identify and quantify the magnitude of the phase change between current and voltage. If a circuit was purely resistive, no phase change between current and voltage would occur. This would hold true for both DC and AC signals. However, no such circuit exists. All circuits possess the properties of resistance, capacitance, and inductance. Phase changes always occur when AC signals are present.

The actual phase displacement can be calculated but the derivation will require calculus which is beyond the scope of this book. What is important to understand is that, while phase changes always take place, they are only significant when frequency roll offs occur.

All circuits have a frequency bandwidth. Amplifiers have a bandwidth and the subcircuits inside the amplifier have a bandwidth. When the bandwidth is exceeded, the audio signal becomes attenuated. The farther in frequency past the beginning of attenuation, the more the signal is reduced. We say that the amplifier "rolls off." Signal roll off points *almost always* exist at both high and low frequencies. I say almost because some amplifiers are flat to DC and experience no lower frequency roll off. (This is not necessarily an advantage.) By manipulating where circuits roll off, another kind of device can be constructed called a *filter*. Filters are used in speaker crossovers, tone controls, equalizers, and any kind of analog signal processing.

There are basically three types of passive filters commonly referred to as high pass, low pass and bandpass. Filters are also designated as first order, second order, third order, etc. The order of a filter determines how quickly it attenuates a signal. The units used to express these rates of attenuation are decibels (dB).

THE DECIBEL

The decibel is one of the most widely used parameters in audio. The term means different things, depending on the application. It is actually a dimensionless quantity. It is a ratio. Its purpose is to compare two different levels. The obvious question is why not just compare the two levels and skip the dB transformation?

There are many processes in nature that occur over incredibly huge ranges. The eyes operate over a range of a tiny fraction of a foot candle at night to tens of thousands of foot candles in bright sunlight. The ears operate over a range of one million billion from the softest sounds to the loudest. Furthermore, perceived changes in levels are based upon percentage increases not absolute increases. To your eyes, the difference between one foot candle and ten foot candles is enormous. The difference between 1000 foot candles and 1010 foot candles is not detectable. The intensity of light will have to increase to 2000 foot candles before most people can detect an increase of brightness. The ears work in a similar fashion. The decibel is a mathematical function that compresses these ratios into percentages that more closely resemble natural experiences.

The absolute minimum level of sound that the human ear is supposed to be able to detect is 0 dB. There is a specific pressure associated with that quantity. The absolute minimum resolution in sound intensity that a highly trained ear is capable of detecting is 1 dB. The average untrained person is supposed to be able to hear a change of 3 dB. A 1 dB increase in loudness requires 10% more amplifier power and a 3 dB increase requires 100% more amplifier power. This means that for the average person, an amp putting out 50 watts doesn't sound any louder until it outputs 100 watts! For sound to be perceived as sounding twice as loud, an amplifier is required to increase its output by 10 times, or 1000%, an increase of 10 dB. As you can see, the ear is very insensitive to changes in volume. This is because it has to function over such an incredibly huge dynamic range.

The application of decibels relating to amplifier power and apparent loudness are a 10-log function. (Don't worry about the math.) When the decibel is applied to voltage gain, it is a 20-log function. That means if you double the voltage, the decibels increase by 6 dB. Similarly, if you reduce the voltage gain by a factor of two, it decreases by 6 dB. This is double the amount just stated for power. A voltage gain of 10 times is 20 dB, and 100 times is 40-db. Just remember to always differentiate between power and voltage gain when using decibels.

When two voltage amplifying components are connected together, the total voltage gain is the sum of the two. A power amp that has a gain of 20 dB and preamp that has a gain of 20 dB combine for a total gain of 40 dB. Ten times ten is 100 and 40 dB is a gain of 100 times. Components that introduce losses are calculated the same way except the negative gain is subtracted.

FILTERS

A low pass filter is one that lets frequencies below a certain threshold pass and attenuates higher frequency signals. A typical application would be a crossover for a woofer. A 1000 Hz low pass filter would pass signals less than 1000 Hz with little or no attenuation but reduce signals above 1000 Hz. The transition between signals that are and are not attenuated isn't abrupt but is smooth and continuous. The rate of attenuation starts off very slowly, long before the "corner" frequency of 1000 Hz, and then increases until a constant rate of attenuation is obtained. The ultimate rate of attenuation is determined by the order of the filter.

A first order filter has an ultimate attenuation rate of 6 dB per octave. That means when the frequency is doubled from the corner frequency, the signal has been reduced by 6 dB or cut in half. The signal at the corner is down about 3 dB. In our example of a 1000-Hz low pass filter, the signal would be down about 1 dB at 500 Hz, 3 dB at 1000 Hz, and 6 dB at 2000 Hz. At 4000 Hz the signal would be reduced by 12 dB or 75%. By 8000 Hz the signal would be attenuated by 18 dB or 87%. The frequency of 8000 Hz is three octaves above 1000 Hz, hence an attenuation of 18 dB.

A second order filter has double the attenuation rate of the first order type. Using an example of a 1000-Hz second order low pass filter, the attenuation at the corner frequency would be 6 dB. At the first octave, the signal would be reduced by 12 dB. By the time the third octave is reached at 8000 Hz, the signal would be reduced by 36 dB or only 1.6% of its original level. This is a very rapid reduction in signal strength. Higher order filters have even greater rates of attenuation.

A high pass filter is just the inverse of the low pass type. It passes frequencies above the corner, and attenuates those at lower frequencies. A typical application would be a crossover for a tweeter.

Connecting a high pass and low pass filter together forms a bandpass filter. This type of filter passes a group of frequencies between the high and low corner frequencies and attenuates all others. A typical application for this type of filter would be for a midrange driver.

By incorporating filters with active components, all manner of wave-shaping circuits can be produced. Tone controls utilize high and low pass filters that boost and cut signals above and below a specific frequency. Bandpass filters are used for equalizers as they partition the audio spectrum into sections that can be individually adjusted. Filters are indispensable in the production of recorded music as well as in playback systems.

REAL VERSUS REACTIVE POWER

This subject is the source of much consternation among engineering students, including yours truly. It is traditionally taught with strange mathematics that use concepts called "complex numbers" that use indefinable quantities like the square root of negative one, which does not exist, so they call it "j" and use it anyway. Fortunately we don't have to get into that, but it is important that the concept of *reactive power* is differentiated from *real power*, because it has everything to do with driving a loudspeaker.

When a resistor gives off heat, that is real power. When a motor shaft rotates a machine and does mechanical work, that is real power. The consumption and dissipation of real power does not involve a phase shift between current and voltage. It is therefore, consumed by resistive elements. The units of real power are *watts*, like a 100-watt light bulb.

The existence of inductance and capacitance cause the phase between the voltage and current of a signal to decouple and become displaced from each other in time. The time displacement requires a special type of power to satisfy it called *reactive power*. The units of reactive power are *volt-amps or vars* which stands for volt-amps reactive. Whatever device is powering the load must supply sufficient reactive power to satisfy the demands placed on it by the load. This is always true whether the device is a utility generator lighting a city or an audio amplifier driving a loudspeaker.

Now here's the tricky part. The classical way this is taught is that reactive power is referred to as *imaginary* power. That's because imaginary numbers are used to calculate it although it really does exist. The difference between reactive power and real power is that all real power is eventually dissipated as heat. Reactive power is *never* dissipated as heat. Since it is not dissipated, it is said to be imaginary, but it really isn't. Are you confused yet?

What is important for us to understand about this subject is that loudspeakers are often highly reactive loads that require reactive power in order to operate. Often times, impedance plots of speakers are published. They describe the impedance of the speaker at any particular frequency. This does not tell the whole story. If a speaker has a stated impedance at a certain frequency of 3 ohms, the label "3 ohms" does not reveal what percentage of the impedance is real or reactive. It could be almost entirely resistive at that frequency, which would require mostly real power, or it could be almost entirely reactive, which would require mostly reactive power. Or, it could be any combination between. There is a way of documenting these ratios using a polar impedance plot which I've always found difficult to read. We should all acquire familiarity with these plots when they are available.

Basically, the more reactive power the amplifier must provide for the speaker, the more difficult the speaker is to drive. Amplifiers are commonly tested with purely resistive loads. This is done for uniformity but does not correlate to real-world conditions. The danger for the consumer is that an amp that performs well into a resistive load may fall apart into a highly reactive load. Lower power amps tend to be more susceptible to these problems. All kinds of things can go wrong, like premature clipping, excessive distortion, ringing, and just sheer bad sound. It would be beneficial to all if power amplifiers were tested into a "standard" reactive load to more closely simulate actual conditions. Getting the industry to agree on a universal standard is probably an impossibility. My point is that reactive testing of amplifiers is just as important, if not more so, than reactive testing of loudspeakers. The speaker generates the need for reactive power, but the amplifier must provide it.

INTERCONNECTS AND GROUNDING

Over the many decades of the audio industry, several standard configurations have evolved to connect two pieces of audio equipment together. Each is valid for its intended application. Unfortunately, the reasons for using one or the other are often obscured to the audiophile because they are constantly being bombarded by advertising claims of the best or ultimate this or that. Let's see if I can bring some clarity to the issue.

The most common type of audio interconnect is the coaxial cable type. It consists of a central conductor surrounded by a thick insulation which in turn is covered by a shield and an outer plastic jacket. The audio is sent down the central conductor and the return path is the shield which is connected to ground. The shield functions as both a current carrying conductor and an extension of the chassis which shields the central conductor from interference from external electric fields. This is the classic unbalanced audio cable.

The coaxial cable contains both capacitance and inductance. Due to the concentric conductors, a magnetic field develops when current flows through the cable, the amount of which is very small and is only significant at radio frequencies. Television antenna cable employs the use of this inductance which is engaged to develop common 75 ohm coaxial antenna cable. The coaxial structure is also two electrodes separated by a dielectric, which forms a capacitor. Cable capacitance is much more dominant in an audio system and must certainly be compensated for.

Cables must do two things. One is to get the signal from here to there. The other is to keep it from being polluted with various types of hum and noise. The biggest offender is power line noise and its harmonics because we are all immersed in it. Even though the cable is shielded, noise can still get in. How does this happen?

The analysis involves studying the interactions of power transformers and power line grounds. Just what is that third prong in the outlet for anyway?

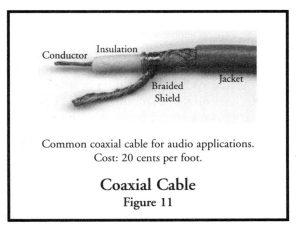

Common coaxial cable for audio applications.
Cost: 20 cents per foot.

Coaxial Cable
Figure 11

In an electric power circuit, there are three wires: hot, neutral and ground. The hot (black) is held at 120 volts and is the source of power. The neutral (white) is held at zero volts and functions as the return path for the current. The ground is also held at zero volts but it is not intended to carry any current. It is in fact tied to the neutral bus in the service panel but has a completely different function, which is to keep people alive. For instance, if an electric dryer developed a short circuit and caused 120 volts to come in contact with the chassis, what would happen if it was not grounded? Let's say the dryer is in the basement and that it is mounted on rubber feet. The short occurs which applies 120 volts to the body of the appliance. You come along in your bare feet, standing on the damp concrete floor, with wet hands from pulling clothes out of the washer. As soon as you touch the dryer, you die. Get the picture?

The circuit breaker did not trip because the chassis was not grounded. It was *floating* with respect to ground. There was no return path into which the fault current could flow. When you touched the dryer, your body provided a path to ground for the 120 volts to flow through, hence your sudden electrocution.

A properly grounded system would have the dryer chassis solidly connected to a grounded conductor. When the short occurred, a heavy fault current would have flowed, which would have immediately tripped the circuit breaker, thereby disconnecting the power. It's not a good idea to leave a shorted chassis energized.

This is the function of the third prong. It keeps the body of any appliance from acquiring a lethal voltage. Nowadays, most appliances are double insulated. They are made out of plastic so there are no conductive parts with which to come into contact and they do not require a three-wire ground type power cord.

High end stereo equipment is not double insulated. It is made of metal. Tube equipment has very high DC voltages inside it. I take no chances; all of my equipment is solidly grounded and provided with a three-wire ground power cord. Never defeat that third prong!

Just as grounding provides for electrical safety, it creates a baseline to which all audio signals can be referenced. In an improperly grounded audio system, voltages can be impressed on individual chassis which cause them to float to different potentials causing all kinds of problems with noise. What is the source of these voltages?

The culprits that cause these voltages are power transformers. The configuration of the interconnect cables has nothing to do with it. They are merely the recipients of the problem.

There is capacitance between the primary and secondary windings of any power transformer. This provides a pathway for coupling the AC power line directly to the chassis. The flow of current is small, on the order of a few milliamps or less, but it is large enough to cause problems. Since the chassis is grounded to the third prong, a pathway is now in place for this leakage current to flow. As it works its way to earth ground, it causes a small voltage drop between the equipment ground and earth ground. The voltage drop doesn't have any effect on the equipment causing the leakage current. Problems arise when a cable connects the offending equipment to another device. The connected device has a different transformer causing different leakage currents. The respective voltage drop of the second equipment item from chassis to earth ground is then different. The two chassis grounds are connected together through the cable shield. Since the grounds are at different voltages, a current will flow in the shield. The voltage difference between the two grounds is added in series to the audio signal and we now have a classic ground loop with associated hum and noise. In addition to the 60 Hz fundamental, power line harmonics also get through and cause buzzing sounds.

Of course, this scenario doesn't always happen but it could and does occur from time to time. It is most prevalent in professional sound reinforcement applications because cable lengths often run into the hundreds if not thousands of feet.

Another problem caused by transformer coupling capacitance is that it can couple rectifier spikes back into the power line and circuit ground. If present, these spikes can cause buzzing sounds. Again, these problems are application specific and you needn't be concerned with them unless you are experiencing stubborn noise problems.

What can the audiophile do to minimize problems with ground loop hum? The choice of power transformer is the first step. Toroid transformers have very low interwinding capacitance, greatly minimizing leakage currents. This stops the problem before it starts. A conventional stacked core transformer can be made with an electrostatic shield installed between the primary and secondary, called an isolation transformer, which also greatly reduces these currents. Another tactic is to properly design the power supply so it doesn't feed noise back into the primary and circuit ground. Most designers are concerned about noise entering the power supply from the power line but the opposite can occur when noise is generated by the equipment's power supply which ultimately gets into the circuit ground.

One could use an interconnect with a heavy shield conductor to minimize any voltage drop through it. Also, one could use short interconnects for the same reason. Plug all the components into one outlet strip and then plug the outlet strip into a grounded outlet. This helps tie all the grounds together and equalize any voltage differences. These techniques should stop any serious ground loop problems.

The established practice in audio land is to use short speaker cables and long interconnects. Once again, I do just the opposite. The audio signal is subject to interference and noise pick up when it is at low levels and loaded into high impedances. Any noise that gets into the input of an amplifier will be magnified by the amp's gain. The problem is worse in preamp inputs because there is more gain between the noise and the speaker, amplifying it further. Long interconnects necessarily present a lot of capacitance to the signal source which is difficult for them to drive because of their limited current capability. An amplifier output operates at an extremely low impedance and there is no gain between it and the speaker, making it virtually immune to noise pick up. Power amps have high current capability which can easily drive any cable capacitance. A properly sized speaker cable will introduce less than 50 milliohms of resistance, which is entirely negligible. I say keep the interconnects short and put the length in the speaker cables.

Another type of cable is the twisted shielded pair. It was originally designed for balanced lines but can also be used for unbalanced applications. It has two conductors and an overall shield. When used in an unbalanced configuration, one conductor carries the audio

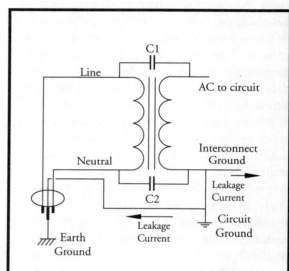

Capacitors C1 and C2 represent interwinding coupling capacitance and provide a pathway for leakage current. The currents travel through the power cord ground and the interconnect ground. They also couple the power line to the circuit allowing high frequency transients to pass through the transformer. The leakage currents produce a voltage drop that causes the circuit ground to be at a slightly higher voltage than earth ground. If excessive, this voltage will cause hum to be introduced into the system. Toroid transformers greatly reduce these effects.

Transformer Coupling Capacitance
Figure 12

signal and the other provides a return path which is connected to circuit ground. The shield is also connected to circuit ground, but at one end only. The shield should always be connected at the signal source end. The inner two conductors are twisted together which helps reject interference from magnetic fields. The shield provides electrostatic shielding from electric fields. By connecting it at one end only, no current can flow through the shield. At the same time, it keeps any electrical fields from causing interference for either the signal or ground conductor. If current was allowed to flow through the shield, it could create a ground loop.

Notice that I am making a distinction between the shielding of magnetic and electric fields. Shield material is nonferrous and can only stop electric fields. Magnetic fields can penetrate a nonferrous shield and induce a current in the underlying conductors. Remember magnetic fields have a directional component. When the inner conductors are twisted together, the magnetic field hits opposite sides of the wires each time they twist around and any induced currents mostly cancel out. The benefits of magnetic field rejection can only be realized if there is a field present that is causing problems. The benefits of the open-ended shield have more frequent applications and could be the answer to stubborn noise problems.

The shielded twisted pair is inherently superior in its ability to get an audio signal from here to there without picking up any noise. This doesn't mean that the coaxial type of audio cable should not be used. Either can provide excellent results depending on the environment in which it is used. If noise pick up is a problem, or the interconnects are more than a few feet in length, then give the twisted shielded pair a close look.

There are many interconnects sold in today's market that have no shielding. They are merely two or three small conductors twisted together and covered with a fiberglass sleeve. Some have their conductors braided together. Many believe that they have a superior sound to the shielded type. A necessary consequence of this arrangement is that these cables are highly susceptible to interference from electric fields. That means they easily pick up hum and noise,

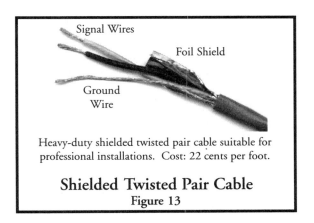

Heavy-duty shielded twisted pair cable suitable for professional installations. Cost: 22 cents per foot.

Shielded Twisted Pair Cable
Figure 13

which is extremely unsatisfactory. If eliminating the shield improves performance, this indicates that the signal source has not been properly designed to drive a reasonable amount of cable capacitance. It should not be necessary to sacrifice shielding for performance.

As an experiment, I encourage the reader to buy a 100-foot spool of 22 gauge stranded wire (less than $10) and twist or braid your own interconnect cables. Once you get going, it shouldn't take more than a couple of hours to make two one-meter cables. The fiberglass sleeving may set you back another $10. Parts distributors also sell high quality RCA connectors for about $5 to $7 each. The entire project will then cost about $45. Assembly will be much easier if you buy two or three different colors of wire for quick identification but that will add another $15 to $20 in cost. With three conductors, you can experiment with using one as a shield by leaving one end open and ground the other, just as is done with a twisted shielded pair. If you like the way it sounds, the project was successful and you just saved yourself a considerable amount of money.

Likewise, there are many professional audio cables that can be used for interconnects and purchased in the required lengths for about $2 per meter. Solder the connectors on the ends and you're done.

BALANCED LINES

Balanced lines are one of the most misunderstood topics in audio. Many think they are absolutely necessary. Oftentimes, people think they have them but really do not. Most do not have any idea how they work but are convinced of their superiority because there is something psychologically comforting about having something that is symmetrical and *balanced*. You don't suppose the mass marketers have been exploiting our cultural desire for symmetry, do you? Let's start at the beginning and discuss where they came from and what their intended uses are.

More than 100 years ago, before the age of vacuum tube amplifiers, the telephone company had a big problem trying to send voice signals miles away. Operating on just the power of the microphone, how could that feeble signal be transmitted many miles down the road without any means to boost it? The signal cables were routed on the same poles as the power lines, immersing them in strong electric fields. Dealing with signal loss is bad enough, but what about all that noise from the power lines that would invariably swamp out the signal? With some brilliant engineering, the phone company invented the balanced line. This was a major milestone in the evolution of telecommunications technology. What they came up with was two number 6 wires separated by 12 inches. It had a characteristic impedance of 600 ohms. How did the balanced line solve the noise problem?

All voltages have to be referenced to something. That's why a voltmeter has two probes. Whatever the negative probe is in contact with forms the reference for the measurement. The voltage that the meter measures is always the *difference* between the probes. It doesn't care about any other voltages in the universe–just the voltage between its probes. If the negative probe is connected to a place in a circuit that is 100 volts above ground, and the positive lead is connected to a point that is 110 volts above ground, the meter just responds to the difference of 10 volts.

In a balanced line, the audio signal is transmitted as two equal and out of phase signals. The signal voltage between the two lines is twice the voltage to ground of either line. The power feed to an electric dryer is an example of a balanced line. The voltage between the two power conductors is 240 volts. The voltage to ground of either power conductor is 120 volts. The two power signals are therefore balanced over ground.

The benefit of this arrangement is that the desired signal is only expressed as the voltage *difference* between the two cables. If a balanced line is immersed in an electric field, the same noise signal will be impressed on each cable. The noise signals are then equal to and in phase with each other. The voltage difference between any two inphase signals is zero. Thus, by responding to only the voltage difference between the two cables, the noise that is common to them (common mode noise) is ignored. A balanced system can extract a useful signal even if the noise signals are greater in amplitude than the desired one.

It is important to note that it is critical for effective noise rejection that the impedances of the signal driver and receiver be well balanced. Noise rejection will suffer greatly if an exact balance in the line's terminating equipment is not maintained.

All of this sounds wonderful, but what does it have to do with audio? That depends on the application. In professional sound systems used in arenas and stadiums, it is common for audio cables to have lengths of several thousand feet! That's a long way to carry a one volt signal. Furthermore, the sending and receiving equipment is probably connected to different power systems with different ground potentials. Under these circumstances, balanced lines are a must. Oftentimes, because of the problems with unequal ground voltages, the lines must be

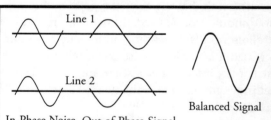

Lines 1 and 2 represent the two conductors of a balanced line. In-phase noise is equal and common to both lines. Out-of-phase signals are the desired audio. The voltage between the two lines is the balanced signal which has twice the amplitude of the out-of-phase signals. The noise does not appear in the balanced signal because the voltage difference between the two in-phase signals is zero.

Balanced Line Noise Rejection

Figure 14

isolated with transformers to decouple them from ground voltage differences.

How does this apply to a residential based, consumer audio system? It doesn't. The practical requirements necessitating balanced lines do not exist in a typical residential audio system. The cable lengths are just too short to matter. If you want to install your power amp two blocks down the street, then by all means use balanced lines. If all your equipment is in the same room, then there is nothing beneficial to be gained. It is not possible to eliminate a problem if the problem doesn't exist in the first place.

I can feel the anger swelling among the devotees of balanced lines. It has been my experience, when dealing with customers, that followers of the balanced line faction tend to be a rather fanatical bunch. No matter how hard I try, I cannot convince them of the soundness of the argument presented above. Their convictions are so strong that some insist I install XLR jacks on my amps for them. I tell them I can do that but the input still isn't balanced. It is just an XLR jack configured an a single ended RCA jack. They still don't care. What I then do is connect pin one to ground, put a dummy load resistor on one signal lead and connect the other signal lead into the circuit. The customer sees the XLR connector and is happy. He connects his balanced line cables and all is well, even though the balanced cable is functioning as a single ended cable by virtue of the connection at the amplifier. The consumer should exercise caution when XLR jacks are installed in audio equipment. Their presence does not guarantee a balanced input. It is highly possible that they are just a different type of receptacle for a single-ended signal. At least I tell my customers what they are going to get. How they accept the information is out of my control.

True balanced line operation can only be achieved if the input port is a truly balanced input, and that requires special circuitry. The only other way to get a truly balanced input is to use line input transformers.

Recently, I built an OTL for a customer who had a requirement for a balanced input. I installed a pair of high quality line input transformers to accommodate him. The amp's circuitry was still unbalanced. The interconnect and signal source that I used were also unbalanced. I am happy to report that the sound quality was excellent. The transformers did not degrade the sound in the least. The transformer input floats the interconnect ground from the chassis ground. They also incorporate an electrostatic shield between the primary and secondary windings. These features serve to isolate the input from any rectifier spikes coupled through the power transformer windings. An electronic balanced input cannot do this. Quality input transformers are expensive and cost about $140 per pair. However, they certainly don't hinder performance and may be the perfect fix for stubborn hum and noise problems caused by leaky power transformers or other system problems.

TRANSMISSION LINES

A common marketing tactic used in audio is to apply real engineering concepts from areas that have no bearing or connection to a residential based audio system. The products claim to eliminate various problems while in reality, the problems were never there in the first place. Another tactic is to refer to properties that do exist, but in amounts so small as to be totally irrelevant. The units of measurement for these events can be one billionth of the smallest increment that can be detected. Nonetheless, they are quoted as being highly significant problems. An example would be a ten-foot long structural I-beam that is designed to carry a load of 20 tons, concentrated in the middle. A fly landing on the center of the beam will cause it to deflect. The amount of movement may be less than the diameter of an electron, but it still moves. The deflection caused by the load of the fly is insignificant and will certainly be omitted from any structural engineer's calculations. An area in audio that often uses irrelevant science in a similar fashion is high frequency radio transmission. The subject usually targeted is transmission lines.

Transmission lines are necessary when sending audio signals over tremendous distances, like 50 miles! They are also required for transmitting high frequency radio signals. What do these two applications have in common?

The factors that mandate the use of a transmission line are the length of the line versus the frequency of the transmitted signal. Frequency determines wavelength which is the actual physical distance a sine wave travels before it repeats itself. Wavelength is then calculated by dividing the speed of wave motion by the frequency. A 500 Hz audio signal traveling at light speed has a wavelength of 372 miles. A 20 kHz signal has a wavelength of 9.3 miles. A 100 MHz radio signal has a wavelength of 9.8 feet. When the length of the line is a significant portion of the signal wavelength, unusual things happen. The signal can reflect back and forth from the ends of the cable and cause interference with itself. So much interference can be generated that very little power flows down the cable. This is not a good situation if you are trying to transmit a radio signal because the signal transmission will be very weak. In radio talk, a measurement of this parameter is called the standing wave ratio, or SWR.

A common TV antenna installation will acquaint you with the procedures for minimizing this problem. The antenna has a transformer output with a 75 ohm impedance. The coaxial cable has an impedance of 75 ohms and the antennae input on the television is rated at 75 ohms. All of the impedances in the system must be equal.

A transmission line can be configured as a coaxial cable or a balanced line such as originally used by the phone company. The terminating equipment is what makes it function as a transmission line, not the cable itself. Cable impedance is critical, but without the proper hardware at each end, the transmission line will fail to work.

The cable lengths in a home-based audio system rarely exceed 15 feet in length. As can be plainly seen, this is such a tiny fraction of the signal wavelength, that transmission line effects do not occur. Any comparisons made to radio frequency phenomena are therefore completely invalid.

BI-WIRING SPEAKER CABLES

When I first became involved with the retail end of audio, I was immediately introduced to the concept of bi-wiring loudspeakers. This is where two separate speaker cables are used, one for the high frequencies and one for the low. My initial reaction to this was total bewilderment. What's this guy doing? It was explained to me that bi-wiring separates the high and low frequency signals on the way to the speaker which improves fidelity. I had to sit down after that one.

Let's look at this objectively. When a composite audio signal travels through an amplifier, it moves as a complete unit. It is not split up by frequency bands. On its journey, all kinds of things happen to it. The amplifier pushes and pulls, distorts it, adds noise and shifts its harmonic structure around. Now are we supposed to believe that after the signal leaves the amplifier, it is so delicate that it can no longer travel down one cable? If that is true, why aren't interconnect cables bi-wired. This reasoning alone should cast enough doubt on the issue. In fact, the validity of the concept is totally false to begin with.

We have already determined that it is the electric field that creates the signal. When two speaker cables are connected to the output of a power amp, the electric field representing the audio signal is equal and present in both. There can be no reduction in interference between the high and low frequencies because they are both present in each cable. The crossover in the speaker still has to filter out the unwanted frequencies when using two cables just as when using one.

If two cables are used, the reactive load placed on the amplifier is increased. Resistance is reduced as a parallel feed is established, but it was probably already so low that further reductions would be insignificant. The reduction in resistance cannot improve the damping factor because the second cable is not connected to the woofer. The load placed on the amplifier is changed but the signals driving the woofer and tweeter are not altered. When one cable is used, the woofer and tweeter "see" the reactance of one cable between themselves and the amplifier. If two cables are used, the woofer and tweeter still "see" the same reactance of one cable between themselves and the amplifier. The same signal travels through both cables. The reactance of the cable is what alters the signal and, in each case, the reactance does not change. Therefore the signal does not change. The current flows in

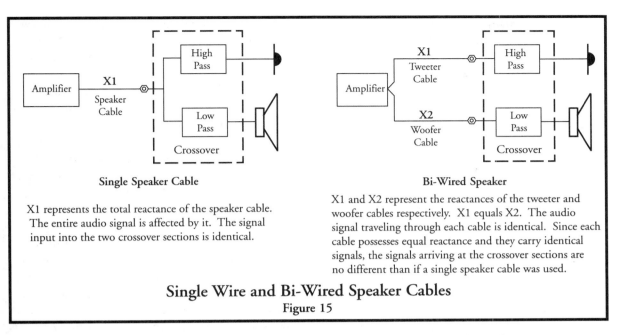

Single Wire and Bi-Wired Speaker Cables
Figure 15

X1 represents the total reactance of the speaker cable. The entire audio signal is affected by it. The signal input into the two crossover sections is identical.

X1 and X2 represent the reactances of the tweeter and woofer cables respectively. X1 equals X2. The audio signal traveling through each cable is identical. Since each cable possesses equal reactance and they carry identical signals, the signals arriving at the crossover sections are no different than if a single speaker cable was used.

the two cables are different because they have different loads. The only mechanism that I can identify which could possibly be attributed to changing the sound is that the combined current when using one cable could create more skin effect. However, the crossover should separate these factors out as skin effect is frequency dependent. The net result is that the tweeter and woofer should both recieve the same signal with one or two cables. It appears that the primary beneficiaries of bi-wiring are most likely the people selling the cables.

POWER SUPPLY EQUIPMENT

This is another subject that is not going to increase my fan appreciation rating. I sure am glad I'm not running for public office. Dealing head-on with reality alienates each faction one by one until everyone is mad at you. That's why politicians never give a straight answer about anything. I just don't have the right stuff for that kind of behavior.

What does a power supply do? What is its function? A power supply in any piece of equipment takes the raw AC power from the wall outlet and converts it into DC at the voltages necessary to operate the circuitry. The power supply contains devices that filter out noise, transients and various electrical garbage from the power line to create a clean source of power for the equipment. A properly designed power supply provides excellent isolation from the outside world and allows the equipment to exist in its own little power universe.

POWER CORDS

What then is the function of the power cord? AC power is an electrical signal. All of the previous discussions about transmitting audio signals and connections apply to power cords. The discipline that specifically deals with the transmission of electrical power is called power engineering. I worked in this field for 16 years as a power systems engineer. A basic discussion of power transmission characteristics is necessary before power cords can be investigated.

All AC power distribution systems possess an impedance. The impedance is different at every point within the distribution system. Looking into any outlet, the net impedance at that point is primarily controlled by all the wiring on that circuit, the appliances operating at that time in the house, the power panel, the cables to the power pole, the utility step down transformer on the pole, and all the other houses connected to the utility transformer. The calculations can be taken back further, but these are the major factors. The impedance of the power cord is in series with the net contribution of all these elements. What do you suppose is the ratio of the impedance of the power cord to the impedance of the power system? It is insignificantly small. It can be deduced then that the impedance of a power cord is so small compared to the power system that it can have no material effect on the delivery of electric power. Having said that, let's assume that power cords do make a difference in audio performance. Is there any other possible mechanism involved? Yes there is. Power cords carry the earth ground cable. It may be possible that different power cords have different impedances in their ground paths that could have some effect on audio performance.

We discussed that transformers have leakage currents that flow through the ground system. Perhaps there is some connection between those leakage currents and ground impedance in making subtle changes to the sound. There is also a minute amount of leakage current that bleeds into the ground from the two power conductors. This is caused by cable capacitance and is very low, on the order of a few microamps. Some audiophiles have told me that a power cord made the background sound blacker. I will infer that they meant quieter. Either that or their lunch was disagreeing with them. A quieter background could be the result of a superior ground connection particularly in response to high frequency noise. I'm trying to be positive about this.

I have used expensive power cords on my OTLs and could not hear any difference. Some customers have reported the same. My OTL uses a toroid transformer which has very low leakage currents. Perhaps when used with transformers with high leakage currents, a noticeable improvement will occur. This possibility requires additional investigation.

A common claim in advertising touts the benefits of using shielded power cords. The concept espoused is that a shielded power cord keeps nasty AC fields out of your equipment. Now think about this. What do you think happens when you plug a power cord into the back of an amp? The AC line current goes inside! The AC power is cabled to the on/off switch, hopefully through a fuse, and to the power transformer. It's running all over the place! How can shielding the power cord keep AC fields out of the equipment when AC power is being pumped inside the box? The correct answer is the obvious one–it can't. Furthermore, transformers radiate all kinds of AC fields and there is one sitting inside gloriously spewing out AC fields.

There is one possible benefit to using shielded power cords and that involves interconnect cables. Many people use unshielded interconnects which are very susceptible to interference from AC fields. Using shielded power cords should reduce the amount of hum pick up caused by the rats nest of twisted cables behind the equipment rack. Shielded interconnects are basically immune from this type of interference and would not benefit from using shielded power cords.

The consumer concerned about value should check out alternatives. The power cords that I furnish with my equipment can be purchased in single quantities from the parts distributor for $3.98 each. I wouldn't use them if I could detect any degradation. Shielded cords are also available. A Beldon type 17600 is a six-foot, seven-inch, 18 gauge, detachable shielded power cord that has a single quantity price of $6.90. If you want a heaver gauge, a Beldon type 17604 is a 14 gauge version which can be purchased for $10.24 in single quantities. (Source: Allied Electronics catalog no. 980, page 747). Be advised! These cords will not come packaged in a fancy box.

ISOLATION TRANSFORMERS

The origination of isolation transformers is the power industry. They were developed to provide clean, low-noise electric power for sensitive equipment like mainframe computers. An internal electrostatic shield eliminates almost all of the noise from the power line. The shield breaks the capacitive coupling between the primary and secondary windings. They also provide what is called a *derived ground*. This is a clean ground isolated from the power system greatly reducing ground current noise. The transformer creates a separate power system that is essentially decoupled from the building's power system.

What can an isolation transformer do in an audio system? For starters, there is nothing wrong with feeding audio equipment with clean power. Electrically, the inclusion of the isolation transformer can have some interesting effects. Anything powered from the transformer now has a different impedance driving its primary windings. This additional impedance will definitely alter the electrical characteristics of the equipment's internal transformers. As far as the circuits are concerned, they are now being powered by a slightly different transformer. This, by all means, could impact the character of the sound in some subtle way. When working with prototypes, I have noticed a difference in the character of the sound when the product is outfitted with the custom-made production transformer. The magnitude of the hardware change required to generate these results is going from an off-the-shelf stacked core unit to a toroid optimized for the application. Changing the effective impedance of an existing transformer is a much smaller leap, but nevertheless, the possibility exists that a realizable sonic alteration could occur.

The isolation transformer's electrostatic shield will break the capacitive coupling between the primary and secondary windings. This will decouple the audio equipment's leakage paths to the power line and ground. The leakage is initiated by the component's internal transformer which is then interrupted by the shield of the isolation transformer. By doing so, a derived ground is established. The chassis ground is therefore more effectively isolated from the power line. This most definitely could impact the sonic characteristics of the system and could be the dominant factor in any performance enhancements. The magnitude of any changes would be influenced by how much interwinding capacitance existed in the equipment's internal transformer. Consequently, a toroid should be less affected than a conventional stacked core transformer. A value judgement of better or worse will have to be applied by the listener.

Providing a clean, isolated ground could also be beneficial in reducing cable noise and creating a stable ground reference. If the system is already quiet this is a moot point, but a clean isolated ground could be used to help remedy noisy systems. The improved ground reference could help components interface with each other, thereby improving performance.

The isolation transformer could be wired to provide a balanced power system. If the

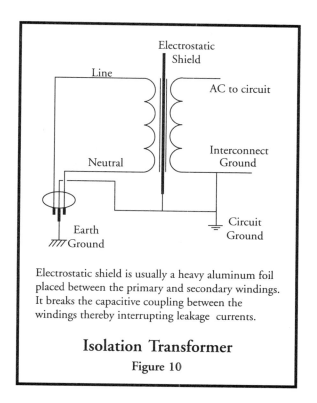

Electrostatic shield is usually a heavy aluminum foil placed between the primary and secondary windings. It breaks the capacitive coupling between the windings thereby interrupting leakage currents.

Isolation Transformer
Figure 10

secondary was center-tapped and grounded, the output would now be 60-0-60 volts, or balanced. This could reduce radiated electrical fields, further reducing noise. Again, any improvement would be dependent upon whether or not there was a noise problem in the first place. Still, it is an interesting concept.

To date, I have logged no experience with isolation transformers in my system and consequently have formed no firm opinions about their effectiveness. There is electrical justification that they can have an impact. Because of the thousands of possibilities of equipment combinations and residence wiring, I would suspect that repeatable results would be difficult to predict.

DAMPING SYSTEMS

I would like to address vibrational damping but have no science to verify or disprove anything. I know of no data that can be correlated to actual results. Does acoustic feedback exist? Certainly it does. How much of a problem it is and where it manifests itself is up for discussion. Having said that, I currently do not use any damping materials in any of my products. Tube type phono preamps are very susceptible to vibration and acoustic feedback whereas power amps are much less influenced by external vibration. This is because power amps have far less voltage gain than phono preamps and the signals through them are 1000 times greater in amplitude. The larger amplitudes tend to override and swamp out any vibrational effects.

As in the case of transmission lines, some of the mass marketers of these products specialize in applying principles of science that do exist. Unfortunately, the magnitudes of these events are so small that they defy all attempts at measurement, just as with our example of the fly on the I-beam. The advertising never makes statements concerning the degree or percentage of the boasted improvement. They just say it sounds better and hand you the bill.

Vacuum tubes are subject to vibrational disturbances. The internal structures can and do vibrate. The exact position of the tube's structures control how they operate. If they move, then the tube outputs a different signal. Therefore, preventing tubes from excessive vibration should be desirable. Note the word "excessive." That's where the ambiguity comes from. How much is too much? It isn't possible to get rid of all of it. From this argument, it seems that techniques that control vibration at the tube would be more successful than ones that dampen vibration at the side panels of the enclosure.

Most tube equipment is either made with many ventilation holes or is completely open. There is nothing protecting the tubes from being directly impacted with airborne sound waves. Powerful sound waves can make furniture vibrate. It would seem reasonable to assume that these same sound waves can make tubes vibrate and there is nothing that can be done to stop them except for putting the amp out in the garage.

Is vibration conducted through the mounting feet much more critical? Should amplifiers hang from rubber straps inside of closets? Or, is the whole endeavor of vibration isolation an exercise in futility? I don't have any concrete answers to give you. My best guess is that vibration isolation products are installation specific. It all depends on the particular situation as to how much benefit will be realized.

There are many products on the market that employ completely different principles, which of course is contradictory. But hey, this is audio. We can't let consistent methodology get in the way of product development. Some systems use spongy, rubbery feet. Some use conical pointed feet of hard material. Some use hydraulically damped feet. Some use heavy platforms of various materials and any type of feet. The types of products range from one extreme to another. I find it comical when reading audio catalogs that sell these products to see how they espouse the wonders of brand X damping system, and then on the very same page, tout the marvels of brand Y, which uses a completely different system that totally contradicts the methods of the tremendous brand X. To say the least, this is very suspicious indeed.

If anyone is interested in investigating this area, go to a sporting goods store and get some small spongy balls and hard rubber balls. Then take a trip to the hardware store and buy some short, fat bolts and grind points on the ends. Then go to a cooking supply store and get a large marble cutting board to use as a platform. Finally, stop by a building supply or landscaping store and buy the mother of all platforms, an 18 inch square by two inch thick concrete paver for a whopping $3.50. Concrete is very brittle, so don't jolt it and make sure it is evenly supported or it will break. The total bill for this shopping spree should be far less than $50 and you now have enough hardware to experiment with for the next six months! Find out for yourself. See what, if anything works. If one particular combination makes an improvement then you're done. Or, you can purchase a commercial product that uses the same general principles. If you know what you are looking for, the odds of a successful purchase will be improved.

Many products employ heavy metal enclosures as a means to dampen vibration. Metal is probably the worst material in the world for that application. Hit it with a hammer and it rings. If it was inherently acoustically damped, the hammer blow would sound like a dull thud. A rubber based material should be much more effective. Loudspeaker cabinets are almost exclusively made of wood products rather than metal for their superior damping capabilities.

The use of heavy metal enclosures brings up the issue of thick face plates. There is absolutely no performance advantage that can be realized from installing a ½-inch thick, machined face plate. These things are horribly expensive and add hundreds of dollars to the retail price. The audio industry has adopted a defacto standard that if equipment doesn't have a thick face plate, it isn't high end quality. This is baloney. The reasoning for their inclusion is that the customer supposedly views them as increasing the "perceived value" of the equipment. I should certainly hope so! The unfortunate customer probably paid $10,000 for the thing. I can "perceive" of a lot of things for that much money and value isn't one of them. As a manufacturer, I refuse to participate in this folly. Perhaps this means that my company does not produce high-end equipment. Instead, I would much rather be known as a company that makes *high performance* equipment at an affordable price. That's what I perceive as *value*.

There is nothing wrong with making stereo equipment attractive. No one wants to own ugly components. However, there *are* ways of making fine looking stereo gear that doesn't double the price. Mainstream high-end manufacturers have chosen to pursue and sell the public on the "look" of massive blocks of machined metal. This industry selected aesthetic necessarily drives up the cost to the consumer so much that it oftentimes doubles the price. The reasoning is simple. General Motors would rather sell 100,000 Cadillacs a year than 100,000 Chevy Cavaliers. Why? Because they make a heck a of a lot more money on each Cadillac. There is much more margin per unit sale. So the audio industry has spent 25 years convincing the buying public that they must have a Cadillac type box to own real high-end equipment. The mass marketers have been so successful with their advertising campaigns that when the consumer sees a preamp advertised for less than $2000, he thinks there's something wrong. The initial reaction is, "It's too cheap to be any good." Prices have escalated to the point that even successful professional people cannot afford new equipment and are forced to buy used gear in the aftermarket.

People in the business have often told me that the mid-priced buying public has abandoned premium audio equipment. I think it's the other way around. The industry has abandoned consumers with incomes of fifty to seventy-five thousand dollars per year and have chosen to only pursue the super rich. Who else would buy a $15,000 CD player? These customers have so much money they don't bother asking the obvious question: What could possibly be inside a CD player to justify a price of $15,000? In situations like this, I feel that the most highly desired feature of the equipment is not the actual performance. It's the price tag.

The structures of transistors and integrated circuits are completely different from tubes. They are solid blocks of semiconductor material and plastic, all molded into one monolithic slab. There are no individual elements that can vibrate. How then can they be disturbed by acoustic feedback? Just because I don't have an answer doesn't mean there is no answer. Then again, I have yet to find any answers from anywhere else.

A loudspeaker is an acoustic transducer. Its purpose is to generate sound waves. A solid-state amplifier does not function as such. Why would principles of vibration isolation intended for a loudspeaker have anything to do with a solid-state power amp? These are two entirely different systems. Even with my tube based OTLs, experiments have been run where sophisticated damping materials were applied to the cabinet and chassis with no noticeable improvement. In any event, the consumer should never install anything inside an amp or on its components that could retard heat rejection, thereby causing failures.

Caution is always warranted when considering these products. This particular field is rife with false science where advertising often quotes scientific principles that do not exist. (This tactic is not exclusive to damping products and appears everywhere.) It's bad enough when totally nonapplicable science is used to sell products, but the mass marketers involved with some of these devices actually *invent* their own science and represent it as fact. Legitimate products are unfortunately thrown in with the mix and, by tolerating the phoney, the credibility of the genuine suffer too. My best advice is that if it sounds bizarre, it probably is.

Audiophiles seem to have an overpowering desire to improve their systems so they try to wring out every last bit of performance. This is the motivation that the mass marketers prey upon to get people to buy their accessory products. If someone has invested $12,000 in a system and finds out about a product that is purported to reap wonderful benefits for $395, the temptation to buy is strong. The "wonder" product may only cost $10 to manufacture and actually does nothing. As competitive as audio is, don't you think manufacturers would use some of these things to gain an edge in the marketplace if they felt they worked? The answer to that question should bring clarity to the issue.

COMPONENT COMPATIBILITY

Every item in an audio system contributes to the final acoustic output. There are no formulas or rules that can be used to determine what a system will ultimately sound like. The acoustics of the listening room play just as important a role. What can be determined is the electrical compatibility between the various components. Understanding how input and output impedances, voltage gain, and speaker characteristics interrelate should make the question of "what works with what" easier to answer.

VOLTAGE DRIVE COMPATIBILITY

The simplest case of component capability is the preamp driving a power amp. The gain of these two components combined with speaker efficiency set the gain of the system. Customers usually ask me what the voltage sensitivity of my amps are when they should be asking me, "What is the gain?"

Unfortunately, there are no standards governing how much signal a CD player or tape deck or any other program source must output. These levels on a CD or phonograph record are set by the recording engineer. Variations in speaker efficiency alone easily traverse a range of 10 dB. The average level of the program source primarily sets the requirement for system gain. That level varies considerably.

Practical experience must therefore be used to establish a baseline for gain. My OTLs have a gain of 20 dB which is a little lower than most. However, many of my customers use passive preamps which have no gain. I am convinced that most systems have too much gain, restricting users from rotating their volume controls past the nine o'clock position. This is a dramatic loss of adjustment resolution. It's like driving a car that goes 10 miles per hour and 35 miles per hour with nothing in between. The setting of the volume control has nothing to do with maximum power output. It is merely an attenuator. Just because there is more rotation does not mean that the amplifier can play louder or that there is more reserve power. Only the useful range of adjustment has been compressed. Ideally, the system gain should be set so that maximum desired loudness is obtained at about three-quarter rotation of the volume control.

As a general rule, when using speakers with an efficiency of 90 to 95 dB, a system gain of 25 to 30 dB should be satisfactory. Speakers with an efficiency of 80 to 85 dB will require a system gain of 30 to 35 dB. Again, room acoustics and listening habits are important factors. These figures are just general guidelines.

The input impedance of power amps can vary from 10k ohms to 100k ohms with most either 50k or 100k. For efficient voltage transfer, the input impedance of the power amp should be at least ten times greater than the output impedance of the preamp. This will result in a signal loss of one dB, which is negligible. The greater the ratio of impedances, the lower the signal loss. Some preamps have output impedances of two or three thousand ohms. When using these products, make sure the selected power amp doesn't have a low impedance input.

Another parameter that must be checked when using high output impedance preamps is the cable capacitance of the interconnect it must drive. If a mismatch occurs, a loss of high frequency response will result. Typically, cables have a capacitance of 30 picofarads per foot, but this specification must be checked for each cable. A three-foot interconnect driven from a preamp with an output impedance of three thousand ohms will create a negligible high frequency roll of 0.3 dB at 20 kHz. At 30 feet, the response will be down 3 dB which could start to affect the treble. Preamplifiers with output impedances of less than 1000 ohms have sufficient current drive to handle the vast majority of installations.

SPEAKER COMPATIBILITY

Electrical compatibility between speakers and amplifiers is much more difficult to establish. The impedance characteristics of loudspeakers vary considerably, making improper matches a common occurrence. Solid-state and tube amplifiers generate current by completely different means. Both must be examined to fully understand the intricacies of speaker compatibility.

Tubes can only pass a limited amount of current. Once that threshold is reached, clipping will occur. Increasing the voltage across the tube will increase current availability, but only minimally. There is just a finite amount of available current.

Transformer-coupled tube amps have different impedance taps on the outputs. When an eight-ohm speaker is connected to the eight-ohm tap, the tubes are loaded with an impedance determined by the transformer. If a four-ohm speaker is connected to the four-ohm tap, the tubes are loaded with exactly the same impedance. As far as the output tubes are concerned, they can't tell the difference between a four-ohm and eight-ohm speaker provided the proper transformer tap is used. Everything is fine so long as load impedance is constant.

What happens if the impedance of the four-ohm speaker drops to two ohms? The impedance presented to the tubes will then drop by a factor of two. Because of the lower impedance, the tubes will have to supply an additional 41% more current to achieve the same power level. They will try to drive the heavy load and will supply a little more current, but nothing near the required 41% will be developed. The result is that maximum power output will fall. Running out of drive current is called *current clipping*. This is a problem all tube amplifiers have with very low impedance speakers. All speakers exhibit variations in impedance. The magnitude of the impedance variations are what cause problems. It is therefore beneficial to any tube amplifier that minimum speaker impedance does not fall much below the rating of the output tap.

Solid-state amplifiers operate in a completely different way. The current capability of a transistor is primarily limited by the voltage of the power supply and the impedance of the load. When a transistor is turned full on, it operates like a switch and will pass as much current as the load requires. If the impedance of the speaker falls by a factor of two, the transistors just pass as much current as necessary to fulfill the load. If too much current is demanded for too long, they will overheat and burn up. Consequently, solid-state amps have over current protection on the output stage to prevent this from happening. The maximum power a solid-state amp can deliver is set by the power supply voltage. Power limits are reached when the amp goes into *voltage clipping*. The output voltage rises and tries to exceed the power supply voltage which of course it cannot do, hence the familiar truncated, or clipped, waveform.

Because of the preponderance of solid-state amplifiers over the last 30 years, speaker designers have not optimized, and many times have ignored, the impedance characteristics of their products. Solid-state amps can easily handle the dips in load impedance but they can be terribly difficult for tube amps. Personally, I feel that this is highly unfortunate and speaker designers should try to minimize any extremely low dips in impedance.

Because of the distribution of musical energy, the impact an impedance dip has on an amplifier is tied to the frequency range in which the dip occurs. Fully 90% of the energy in a typical musical signal lies below 500 Hz. A loudspeaker that may be rated for 100 watts should have a woofer with at least a 100 watt power rating. Its tweeter may only produce frequencies over 5000 Hz and have a power rating of only 25 watts. The reason for the lower power rating of the tweeter is that there simply isn't as much work for it to do. The musical energy is not great enough at those frequencies to necessitate an equivalent power rating. By the time harmonics stretch out to 20 kHz, power requirements may drop to a fraction of a watt. An impedance dip in the power range of 100 to 500 Hz is very significant, while a dip in the midrange rarely presents a problem. Dips at the highest frequencies are of no consequence.

People are constantly asking me about speaker compatibility of my OTLs with electrostatic speakers because they see minimum impedance specifications of 1.5 to 2 ohms for the screens. These lows always occur at 20 kHz and do not represent any kind of drive problem.

I hesitate to mention this, but OTLs are the best amplifiers for electrostatic speakers. The reason for my hesitation is that market perception has labeled my OTL as useful for only this application. This is totally untrue. The amps easily drive dynamic speakers, but market perceptions built up over many years are very hard to overcome. I fear emphasizing the electrostatic application just reinforces these false beliefs. At the same time, I would like to explain why the combination is successful. Electrostatics typically have impedances in the acoustic power range of over 100 ohms. These high impedances require correspondingly high voltages to drive them. My OTLs can deliver up to 70 volts RMS into these types of loads, which is twice what a big solid-state amp can produce. That's why they're such a good match.

VACUUM TUBES

The invention of the vacuum tube is one of the hallmark achievements in the history of technology. Its development marked the beginning of the age of electronics, which probably has been responsible for more changes in human culture than any in history. The vacuum tube's inner workings are worthy of an examination.

A vacuum tube consists of several basic parts. The most obvious is a glass bottle with all the air removed. Electrons don't travel through air very well because air is a dielectric. Vacuum tubes don't have any air in them for that very reason.

Electron motion through tubes is a completely different phenomena than electron motion through a wire. In a tube, electrons actually leave a solid material and literally travel through space. Electric fields are the sources of energy that control the motion of electrons in a tube.

A triode is the simplest type of tube and it has three parts or structures mounted inside the glass bottle. They are the *cathode*, the *grid* and the *plate*. The cathode is a metal sleeve, usually with an oval cross sectional area made of a nickel-based alloy. It is the source of electrons that pass through the tube. The method of releasing these free electrons is heat. The source of the heat is the filament which is usually a tungsten alloy that is closely fitted inside the cathode. It is electrically insulated from the cathode with a ceramic-based material.

When we discussed current flow in a conductor, the random haphazard movement of electrons was caused by heat at room temperature. If a conductor is further heated to a dull red glow, the electrons gain so much energy that they actually leave the base metal and venture out into space. To facilitate the process of electrons leaving the cathode, it is coated with a chemical (usually barium oxide) that dramatically increases this activity.

Once the electrons are jettisoned, they have to go somewhere. As the electrons leave the cathode they gather around the cathode in a type of cloud. Remember, charges with the same polarity repel each other. Eventually the number of electrons becomes sufficient for them to generate an electric field strong enough to repel any more electrons from leaving the cathode. An equilibrium develops where the forces trying to eject electrons equals the forces that are repelling them back. The resultant electron cloud is called the *space charge*.

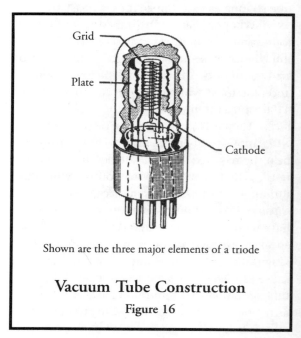

Shown are the three major elements of a triode

Vacuum Tube Construction
Figure 16

Now that there is a quantity of free electrons available in the tube, a way must be implemented to get them to do some work for us. Surrounding the cathode is a concentric cylinder called the plate. A positive voltage is applied to it which sets up an electric field between the plate and cathode. Since the plate is positive and the electrons are negative, the electrons are attracted to the plate. The field applies a force to the electrons and they begin to accelerate toward the plate.

There are no air molecules to impede the flow of electrons. That is, there is nothing for them to bump into. Because the pathway is unobstructed, they continue to accelerate as they travel. By the time they reach the plate, they are traveling at maximum speed with minimum density. The electrons near the cathode are moving at minimum speed and are at maximum density. Therefore the density of electrons varies through the tube.

For every electron that leaves the cathode, it must be replaced. The electric field, set up by the space charge surrounding the cathode, is weakened every time an electron leaves. The reduction in field strength allows another electron to escape the cathode to replace it. The faster the electrons are drawn off, the weaker the field becomes and the more electrons are liberated to replace them. The system constantly balances itself, replenishing electrons as needed.

There is a limit to the rate of electron replenishment. There is only so much cathode material and so much heat. Ultimately, a limit will be reached where no more electrons can be made available. The tube simply runs out of current. This is why tubes can only pass a certain amount of current.

One more element is required to complete the assembly. Surrounding the cathode is a single layer coil of wire with widely spaced turns called the grid. A negative voltage is applied to the grid which has the effect of creating another field between the plate and cathode. The field's negative polarity retards the flow of electrons. It acts as a throttling mechanism that controls or modulates the electron flow. The grid is where the audio signal is applied. The audio then sets up its own field, thereby controlling the tube. The negative voltage on the grid is called the *bias* voltage.

There is another kind of triode where the filament is also the cathode. They are primarily used as output tubes for single-ended amplifiers. The efficiency of electron emission is much lower with this configuration and requires three or four times as much heat. They are harder for the designer to work with but some audiophiles greatly prefer the type of sound they produce.

Another common tube is called the *pentode*. It has a total of five internal structures. One of the two additional structures is the *screen grid*. This is another coil of wire, even more widely spaced than the grid, that is placed between the grid and plate. The electric field created by the screen effectively shields or *screens* the grid and cathode from the plate. The net affect of the screen is to dramatically reduce input capacitance, making the tube easier to drive and increasing gain. The shielding properties of the screen make the tube much less sensitive to plate voltage variations. In fact, plate voltage can often vary a couple of hundred volts before it has any material effect on the current through the tube. Pentodes are much less susceptible to power supply ripple than triodes.

There is a fifth element called the suppressor grid. It is a coil wound between the screen and plate. When the electrons hit the plate, they are going very fast and cause electrons in the plate to be knocked free into space. The suppressor is connected to the cathode so it assumes a negative potential relative to the plate. The free electrons are repelled by the suppressor and find their way back to the plate. This makes the tube more linear and maximizes gain.

Triodes and pentodes have different distortion characteristics. Triodes produce mostly even order harmonics while pentodes produce both even and odd order distortion products. Even order harmonics are generally considered less objectionable than odd order. Pentodes get a bad rap for this as they can still be employed to make great-sounding equipment. Triodes have another advantage in that they are a little quieter that pentodes. The extra grids generate some additional noise. Pentodes can be connected as triodes and in that case, acquire all of the performance advantages of triodes. I often do this.

FEEDBACK AND IMAGING

Now we are going to enter into the treacherous realm of the most notorious scoundrel in all of audio, the domain of the dreaded *negative feedback*. The mere mention of this villain strikes shear terror in the hearts of any high-end audiophile. It has been blamed as the sole cause for every sonic deficiency that has ever existed. Any association with this monster justifies complete rejection and persecution by the true believers. Someone please send me an exorcist to remove this vile corruption from my home!

Of all the concepts that the mass marketers have planted in the minds of audiophiles, none have been more successfully embedded than the virtual brainwashing that has occurred regarding negative feedback. Every belief system needs an enemy to focus upon and negative feedback is the unfortunate target of the wrath of the priests of audio.

Well, I don't think this persecution is at all justified. In fact, it's sheer nonsense. The use of feedback will not condemn anyone to eternal damnation. Please forgive my heresy, for I am but a confused engineer who's been using feedback for years to make some really great-sounding equipment. Perhaps if I was an ordained priest of the high order of audio, my vision would be cleared and I would see the error of my ways and realize that it was the cosmic ray deflector all along. (My analyst says that it's good therapy to get these conflicts out in the open.)

Negative feedback is one of the most useful and versatile tools a designer can utilize to improve the performance of an amplifying circuit. It can reduce noise, lower distortion, extend bandwidth, and reduce output impedance, all at the same time! It can be used to make equalization networks, filters, boost signals and a host of other applications. So what's the problem?

There is a general acoustic correlation to the application of feedback, particularly in power amplifiers. Its presence definitely can alter the acoustic experience. We must define what a proper acoustic experience is before the affects of feedback can be analyzed.

When a recording is made, the usual procedure is to use a multitrack tape recorder and separately record individual voices and instruments. Oftentimes these sources are segregated into separate recording booths for complete acoustic isolation. These individual recordings are called tracks. If several instruments are grouped together, each usually has its own microphone. Rarely is music recorded in concert halls using just a few microphones. The tracks are subject to all sorts of manipulations by the producer and recording engineer. Signals are boosted, cut, reverb is added, and on and on. Elaborate consoles are used, loaded with sophisticated electronics to implement these alterations. Finally, the tracks are mixed together into a two-channel recording. The end product oftentimes bears no resemblance to the original and is the result of the application of very creative technology.

The audiophile buys the recording, plays it on his system and then describes sound staging and imaging. What sound stage? There never was one to begin with.

Psychoacoustic phenomena exist that cause the sensation of sound localization in a two-channel recording. We have all experienced this. Sounds appear to hang in the air, some closer to the left or right, front or back. How does this happen?

Human beings determine the location of sounds by the time delay detected between the two ears. This is a highly precise adaptation which I'm sure was essential to survival in evolutionary times. People had to know where the lion's roars were coming from so they could run the other way!

When music is played on a two-channel stereo system, time-based information is embedded into the signal. All of the sounds and their harmonics start and stop at different times. Time variations obviously also occur between sounds common to both channels. The entire stereo system acts on this information and then adds its own variations on top of those originally in the recording. Furthermore, room acoustics are a major factor in adding yet another layer of directional effects to the sound. The final outcome is a perception in the listener of the placements of instruments. So I have to ask again: What sound stage?

The perception of localization of individual instruments and voices must therefore be an artificial contrivance. This phenomenon is purely a product of the recording process and the playback system. But it's a good thing!

Because of these effects, additional realism is added to the experience of listening to music. Without them, much pleasure would be lost and the music would sound flat and lifeless. The point I am trying to make is that these perceptions are neither correct nor incorrect. They are just there or not, and in differing amounts.

Some amplifiers emphasize localization in different ways, and feedback definitely plays a part in this process. With the addition of feedback, instrument localization tends to be more pronounced and identifiable. With its reduction or elimination, individual sonic images tend to blur and the overall width of the *apparent* sound stage tends to widen.

Identifying and verifying the mechanism that causes this would require years to establish and a four hundred-page book to prove it! I do, however, have an idea as to why this happens.

When feedback is added, bandwidth is extended and rise time is improved. These are changes in the time characteristics of the amplifier. I liken it to focusing a camera. As the image is brought into focus, the individual elements of the picture are brought forth and the resolution is increased. When the image is taken out of focus, the elements blur. The time characteristics of amplifiers can be tested with "impulse" testing using Fourier Analysis. It would be interesting to see if there is a correlation between any repeatable results and the perception of localization. Perhaps this is the wrong avenue to pursue, but it certainly merits further investigation.

For some reason, contemporary audio culture seems to prefer a blurred image with exaggerated width. This effect is enhanced with the elimination of negative feedback. Zero feedback, single-ended amplifiers are quite popular now. Do you suppose the mass marketers have something to do with this?

Getting rid of feedback creates many other problems. Output impedance rises, which ruins the damping capabilities of the amplifier, causing mushy, boomy bass. Noise will go up or expensive power supply filtering will be required to keep it at tolerable levels. Distortion will necessarily rise. The speed of the amplifier will be reduced potentially causing transients to be muffled or lost.

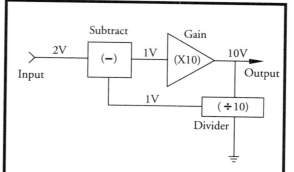

Negative feedback system with a closed loop gain of 5.

Internal gain is 10 times. Input voltage is 2 V. Divider samples output and sends a 1V signal to subtract stage where it is combined with 2V input. Difference is 1V which is fed into gain stage. Feedback reduces gain from 10 times to 5 times by reducing input voltage into gain stage. Benefits are achieved at the expense of gain. All signals are in phase.

Negative Feedback
Figure 18

I don't know about you, but these seem like fairly important parameters to me. The obvious and reasonable solution is to apply feedback as needed to achieve a balance of performance. That's what I do in my designs. Feedback is neither good nor bad, right or wrong. It is simply a tool that can be used to optimize performance.

Electrically speaking, how does feedback work? The mechanism for feedback is rather amazing in that it acts like a computer, but it has no means to think. It is implemented by sampling a portion of the output signal and combining it with the input, usually at the first stage of the amplifier. If an amplifier was perfect and did not alter the signal in any way except to provide gain, the input and output signals would be exact copies of each other, except the output would have greater amplitude. The reason only a portion of the output signal is combined with the input is to compensate for gain. This is done with a resistive divider.

The two combined signals are in phase.

The sampled portion of the output is always lower in amplitude than the input. Because the two combined signals in our perfect amplifier are exact parallels of each other, the result of the combination is a third signal that is also an exact parallel, but has an amplitude which is equal to the difference between the two. This third signal is what the amplifier uses to send to the output. It always has an amplitude less than the original input signal. That's why negative feedback always reduces voltage gain. The feedback effectively reduces the amplitude of the input signal.

Getting back to reality, all amplifiers alter signals in other ways besides increasing amplitude. These alterations could be lumped together in one category and called distortion. When the output and input signals of a real amplifier are combined, the resultant third signal now contains an error signal that has the opposite phase, or inverse, of the distortion. The error signal tries to push the amplifier in the opposite direction that the distorted output signal is going. The two cancel out and the distortion is reduced. The signal traveling through the amplifier circuit will possess additional distortion but it is there to cancel out the effects of the distortion at the output. So an amount of distortion is added to the audio that is the inverse of the distortion naturally produced by the amplifier. The effect of this added distortion is to cancel out and negate distortion in the output. That's what negative feedback does.

SERVO SYSTEMS

These are sometimes used in amplifiers to compensate for DC levels or in subwoofers to linearize cone motion. In an academic sense, they use feedback signals but in a completely different way.

Servo systems are used to control a process. Some kind of output is generated at the end of the process whether it is sound or the position of a mechanical actuator. The first thing the servo does is monitor the output and derive an analog of that output. In feedback, we use the actual output voltage signal. A servo may monitor output current, or second order distortion, or DC output or position of a laser beam in a CD player. The derived analog signal then goes into a process amplifier that may perform a mathematical operation on the signal. The signal could be amplified or scaled or changed in some other way. This is what an analog computer does.

The altered signal then gets injected somewhere in the process causing the desired parameters to change. It functions as a corrective signal. The greater the output is skewed, the greater the magnitude of the corrective signal. It is important to note that the corrective signal does not have to be added to the input, although it can be. It all depends upon what you are trying to do. I might add that servos can and often are implemented entirely with digital microprocessor circuitry.

Servos affect the transient response of a process. In that sense, they share much with negative feedback. Parameters like rise time, damping, and overshoot apply to both. They add complexity to a project but their inclusion opens up many possibilities. I am not currently using any in my products, but who knows what the future will bring?

ACOUSTICS

The whole purpose of processing electrical signals in stereo equipment is to create sound. That is ultimately the final product. There are many similarities in the behavior of sound to electrical signals but there are also many distinct differences that should be discussed.

Sound is a mechanical disturbance propagating through a medium. It cannot be transmitted through a vacuum. Some type of material must be present for it to go through. It can travel through air, water, or a solid. Sound transmitted through the air is actually a pressure wave.

Air is a compressible substance–as anyone who has ever pumped up a tire can attest. Therefore, the pressure wave should not be thought of as air movement. The air molecules do move, but only at microscopic distances and for low level sounds, at atomic distances. Think of the air molecules as just pressing against each other and your ears.

A pressure wave moves through the air just as a ripple moves across a pond. Once a disturbance is set up, the wave front moves out equally in all directions. As the wave travels, its amplitude decreases because the wave's energy is dissipated, or absorbed, into the air. Because the wave spreads out as it moves, its energy is applied over an increasingly larger area. Eventually there is no wave and no sound.

Every sound wave is built of a region of increasing and then decreasing pressure. The "positive" part of the wave is called *compression* and the "negative" part of the wave is called *rarefaction*. (There is no polarity that can be associated with an acoustic wave. I am using the terms "positive" and "negative" here only to correlate with an electrical signal.) During compression, the pressure in the wave is greater than the general pressure in the air. During rarefaction, the wave pressure is lower than the general air pressure. There are three characteristics to the wave that essentially describe its properties. The shape of the wave is determined by the rate at which the pressure increases and decreases. The frequency of the wave is determined by how much time elapses before the pressure goes from compression through rarefaction, and the loudness is determined by the wave's amplitude.

When a loudspeaker cone pushes forward, a compression wave is formed. When it pulls back, a rarefaction wave is produced. The faster the cone moves back and forth, the higher the frequency. The greater distance the cone moves back and forth, the greater the volume. When you place your hand in front of a woofer and feel air movement, that's what it is, air movement, not sound. The air moves simply because the woofer is pulsating back and forth so violently that air is being pushed around. This actually represents inefficiency in coupling the woofer to the air. Energy is wasted in creating a breeze instead of making sound. Most loudspeakers have an efficiency of less than 1%. The most efficient loudspeakers available are not more that about 5% efficient. This means that if an amplifier is producing 100 watts of electrical power, the actual power of the sound projected into the room is probably on the order of ½ watt.

Loudspeaker measurements are often made in an anechoic chamber in order to simulate a "free field" environment. Free field means the sound leaves the speaker and radiates outward with no reflections to interfere with the waves' propagation. This type of sound field is also called a *non-reverberant sound field*. The easiest way for you and I to accomplish this is to set the speaker outside in an open area with no other structures nearby. The anechoic chamber achieves this effect by having special geometries and sound absorbent materials lining all the interior surfaces that capture all of the sound and prohibit any sound from being reflected back into the room.

In a non-reverberant sound field, every time the distance is doubled between the listener and the loudspeaker, the volume, or more accurately the sound pressure level (SPL), decreases by 6 dB. This is because the sound wave is being stretched and spread as it travels. Actually, the reduction SPL has to be a little greater than that because air is not a lossless medium. Because of its compressibility, energy from the wave is absorbed into the air as it travels. This continuously attenuates the amplitude during propagation in addition to losses attributed to wave spreading. The energy content of the sound wave gets soaked up by the air as it travels.

Listening to a speaker in a residence is a totally different experience. This type of environment is called a *reverberant sound field*.

When a sound wave reaches a hard surface, much of the wave is reflected back into the room. The harder and denser the surface, the greater the propensity for reflection. I'm sure you have experienced echo effects in large empty rooms with hard plaster or concrete surfaces. Rooms that are fully carpeted with heavy drapes and much overstuffed furniture, are highly damped and do not tend to exhibit these effects. Soft surfaces absorb sound by letting their molecules vibrate when stimulated by the sound, thereby converting it into heat. Ultimately, all of the electrical energy that flows into a power amp from the wall outlet is converted into heat.

The amount of reflected sound in a typical listening room predominates what we hear. If you stand directly in front of a speaker and slowly move back, you will notice that, at first, the SPL starts to reduce. By the time you have moved back a few feet, the reduction in SPL levels off and becomes relatively constant. This is because there is so much reflected sound energy in the room that it swamps out the effects of the sound wave spreading out as it travels. The loudspeaker doesn't really project a specific type of sound in the room. It stimulates the room with acoustical energy and the room does what it wants with it. The acoustical properties of rooms so dramatically alter the final result, that no speaker will sound the same in two different rooms. This unavoidable fact makes buying speakers the most difficult choice the audiophile faces.

Merely changing the placement of speakers can have a huge effect. Recently, I reversed the orientation of my listening room by just exchanging everything from one side to the other. The room was not symmetrical in this regard. This made a tremendous improvement. Why do these things happen? What mechanism causes the sound to change as you move about the room or move the speakers?

The speed of sound in air is 1087 ft. per second. This is slow enough that many effects similar to those in transmission lines occur. A 100 Hz tone has a wavelength of 10.87 feet. If a room had two reflecting surfaces that were 10.87 feet apart, they would be tuned to harmonics of a 100 Hz signal. As acoustical energy is reflected back and forth, standing waves will form at whole number multiples of 100 Hz. That means frequencies of 200, 300, 400 Hz etc. will be created. These additional frequencies reinforce other signals or reduce them depending upon where the listener is standing. They are of lower amplitude than the fundamental but they are still there. This phenomenon occurs for every pair of reflecting surfaces. If two walls were 15 feet apart then the fundamental frequency would be 72.5 Hz and the generated harmonics would then be 145 Hz, 217 Hz, 290 Hz, etc. There are literally hundreds of frequencies of standing waves causing interference in each listening room and there is absolutely nothing you can do about it.

When sound reflects around the room, it arrives at your ears at different times. This is a kind of echo called *reverberation*. Clap your hands in a highly reverberant room and you hear a pronounced ringing sound. Do the same in a highly damped room and the ring is gone. The ringing sound is caused by sound waves bouncing around until they finally decay. Sound projected into such a room has a lively character to it. Sound projected into a highly damped room may sound dull and lifeless. Neither is correct. Too much of either characteristic can cause excessive colorations.

Other sources of room colorations are mechanical resonances with the structure itself. All structures have natural mechanical resonances, and when they are excited by sound waves, they start vibrating and generate more tones. These resonances can be a real problem in accentuating parts of the music, particularly in the bass region. They can cause a muddy, exaggerated sound at specific bass frequencies. When these resonances are excited, you can actually feel the walls or floor vibrate, often at a considerable level. Years ago when I was living in an apartment, I built a signal generator and was playing around with my system. I went hunting for room resonances and found a major one at around 50 Hz. I remember it was a very sharp, or finely tuned, resonance as the frequency had to be carefully adjusted. I made that apartment shake like you wouldn't believe. The experiment probably upset quite a few roaches!

When sound hits a reflective surface that is larger than one quarter wavelength, it will bounce off it. If the surface is smaller than one quarter wave length it tends to *diffract*, or bend, around the object. This lengthens the pathway that the sound wave can travel. Diffraction, therefore, can increase time delays of reflected sound by increasing the time the sound takes to bounce off something and end up at your ears.

All of the above-mentioned effects culminate in the sound that is heard at any specific location in any room. They all combine in different amounts everywhere in the room causing the room to sound different depending on where the listener is located. Most of these characteristics cannot be altered short of making drastic physical changes to the room's interior. Drapes and carpets can be added as well as other damping and reflecting materials but you are still stuck with the room's basic geometry. Room acoustics are by far the most dominant variable in determining what a stereo system will sound like and, for the most part, is the least alterable by the audiophile.

The science of acoustical treatments is highly developed and widely employed in professional sound reinforcement systems. The methods and techniques used are hardly proprietary to high-end audio. Much literature is available on the subject. Performance characteristics of acoustical damping materials have been widely published in a variety of sources. There are no secret devices and materials. Generally speaking, it takes a lot of acoustical material to significantly alter the sonic characteristics of a room. I have seen the experts in action at the trade shows, and I mean *real* experts. The process of determining "what do you put where and how much do you use" is entirely trial and error. They just keep working with it until they hit on right combination. It's a tough job. The consumer should therefore be highly suspect of any miraculous, instant treatments that are purported to significantly improve the sound of any room.

APPARENT LOUDNESS

Not only is the ear relatively insensitive to changes in loudness, apparent loudness is also heavily influenced by frequency versus relative loudness. Explaining this phenomenon will require the use of a graph. Figure 19 shows a set of equal loudness curves for the human ear. Each curve represents the apparent perceived loudness over a range of frequencies relative to a specific measured loudness at 1000 Hz. This means the listener perceives the sound to be louder or softer than what is heard at 1000 Hz as the frequency changes. Also the curves show how the variation in apparent loudness changes as the level at 1000 Hz is changed.

Looking at the curves, at a SPL of 10 dB, for a 20 Hz tone to sound as loud as it does at 1000 Hz, it would have to measure at 78 db. This is an increase in power of 68 dB or 6.3 million times! At a more realistic level of 70 dB SPL at 1000 Hz, a 20 Hz tone will have to be increased to 108 dB to sound as loud. This is a power increase of 48 dB or 63 thousand times. At a very high level of 100 dB, a 20 Hz tone will have to be increased to 128 dB to sound as loud,

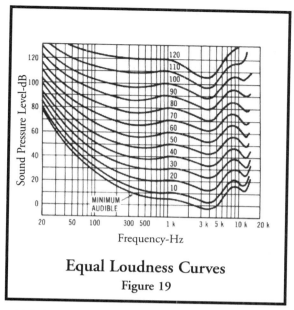

Equal Loudness Curves
Figure 19

which is an increase in power of 631 times. As you can see, producing high power sounds at the lower frequency limit of human hearing takes such colossal amounts of power that is a practical impossibility. Raising the lower limit required of a stereo system to a more realistic 30 Hz reduces the power required from the amplifier by over 100 times. Therefore, the audiophile should not be overly concerned with speakers that roll off at 30 Hz because it is almost impossible to produce sounds below 30 Hz at anything but moderate levels. Unless the audiophile wants to invest in expensive active subwoofers with kilowatt amplifiers, accepting a lower limit of 30 Hz is the only realistic choice. Compact disks are capable of providing sounds well below the limits of human audibility. Just because you *can* do something doesn't mean that you should. There is virtually no musical energy below 30 Hz and very little below 40 Hz. I'm aware that there are a few organs in the world that have a 16 Hz pedal tone and a few drums that generate tones at the very limits of hearing, but that doesn't mean that everyone's system has to be burdened with these very expensive requirements.

Many loudspeaker frequency response tests are run at just one watt. This can be highly misleading in describing the speaker's ability to reproduce bass tones. As has just been discussed, the power requirements placed on the speaker and amp in the lowest octave are sizeable even at moderate listening levels. A speaker that has a flat response curve at a couple of watts may not be able to generate the lowest frequencies at high listening levels. This is a type of dynamic compression. Severe distortion can result when the volume is turned up. Dynamic compression of bass signals explains why some speakers have great bass at low listening levels, but the bottom end disappears when the system is pushed to its limits. The root cause of this effect is the ear's extreme nonlinearity to low frequency signals. Only frequency response tests performed at maximum power will reveal the audible bass performance. It is highly possible that a speaker that rolls off at 35 Hz has more bass than one that rolls off at 25 Hz if the former can generate much more powerful signals at the 40 Hz level.

At the treble end of the spectrum, the curves show us that the ear is most sensitive between 2000 and 3000 Hz. Sensitivity then decreases to about 9000 Hz and becomes a little more sensitive at 12,000 Hz. Because of the very low power content of musical energy in the top octaves, the power requirements of the amplifier and speaker are nowhere near as great as in the bass region.

DYNAMIC RANGE

The ear has a dynamic range of 130 dB. However, the complete range can never be fully realized. First of all, most residential systems are not capable of producing over 110 dB. Getting to 120 dB requires highly efficient horn type speakers, which are available, and many hundreds of watts of amplifier power. It can be done, but is so impractical that it makes this goal virtually unreachable. At the other end of the spectrum, the quietest environment that can be expected in a home is about 30 dB. The sounds of breathing begin to dominate at these levels. It's really hard to listen to music if you can't breathe. Many home environments don't get any quieter than 40 dB. Realistically, the maximum dynamic range that can be achieved is 80 dB and quite possibly only 70 dB. Once the noise floor of a stereo system gets below the noise floor of the listening room, any further improvements are not noticeable because the ambient noise will swamp out and mask the system noise.

MASKING

Masking is an interesting effect that forms the basis for most noise reduction systems. It is a characteristic of human hearing. If two speakers are playing the same tone and one is made just 3 dB louder than the other, the louder speaker will dominate and the tone will sound like it is only being generated by one speaker. The quieter speaker will appear to be off. Masking is then a characteristic of the ear to ignore a sound if there is another sound at exactly the same frequency that is just slightly louder. Perhaps back in evolutionary times, the louder a sound was, the closer and more imminent a potential threat was. Maybe our minds evolved to ignore similar but quieter sounds because the loudest sound probably had the closest teeth.

Audible noise is always present. The only time it is heard is when the program material becomes sufficiently quiet that the noise comes through. At all other times, the program material masks over the noise, rendering it inaudible. Noise reduction systems such as those used for tape decks are prime examples of exploiting this condition. During the recording process, high frequency signals are boosted such that they are well above the noise floor. During playback, the highs are reduced to their original levels. The noise is reduced along with them, causing a significant reduction in background noise. The ratio of boosted highs to noise is preserved because during recording, the noise was not boosted, only the highs. Upon playback after the signal is restored to the original level, the highs mask over the noise because they are still at a much higher level. This description is a simplification. The amount of signal boost is variable depending on the strength of the highs present at any one time.

Another technique that can be used to reduce existing noise is dynamic noise filtering. Back in the 80s there were several products that employed this process. With the advent of the compact disk, however, they have all but disappeared from the scene. They work by actually restricting the bandwidth of the program. When no highs strong enough to mask the noise are present, a variable low pass filter quickly closes down the bandwidth, thereby filtering out the noise. As soon as the music returns, the filter opens up and restores the bandwidth, letting the highs through. The speed of the filter has to be very fast or some of the music will be lost. One of the few remaining applications for a device like this, is to clean up old noisy tapes and records. Otherwise, they have few if any applications in a modern system. I mention them because they exemplify another case of masking effects.

CONCLUSION TO PART ONE

I have talked to so many audiophiles who are terribly disenchanted with the current state of affairs in audio. They often tell me that they have spent many thousands of dollars but the equipment never performs as expected. Quite frequently, mediocre or even poor performance is the result. They are looking for answers to help guide them in their purchasing decisions but are unable to secure reliable information. Ask ten experts and you get ten different answers–depending upon what the experts are selling.

Whenever unrealistic claims are reported in the media, audiophiles tend to develop high expectations that can never be satisfied. I have had many conversations with experienced audiophiles who told me they had purchased various items with high hopes and were terribly disappointed when no appreciable benefit resulted. Their system still sounded basically the same.

Once, an inexperienced audiophile came to my shop because he wanted to hear an OTL. I gave him a demonstration and he was disappointed. He said the sound wasn't "magical" enough. The fellow then examined my setup and informed me the problem was caused by my cables. If I had used the right cables, the system would have possessed the required magical qualities. The demonstration abruptly ended.

This encounter illustrates what inaccurate, exaggerated media reporting can do to the novice. Why did this individual expect magic from my OTL? I certainly didn't tell him that. His anticipation of a supernatural sonic experience was probably formed from reports published in the media. When reporting turns to hyping, everyone loses. The consumer becomes frustrated and confused because he doesn't know what to believe or expect when finally confronted with reality. Most of the time, reality doesn't show up until after the check is signed.

There is nothing inherently wrong with trying to sell products. I am trying to sell *my* products. There *is* something wrong with intentionally misleading the buying public. I am no ultra bleeding heart liberal, but it does upset me to see people throw out their hard-earned money on products that cannot possibly perform as claimed. When shopping at a bazaar, the best advice is "let the buyer beware." Don't totally trust anyone or anything when money is changing hands, including me!

It is my hope that whatever equipment you do decide to purchase, this book will help you get your money's worth and prevent your being taken advantage of by someone. Please don't base your purchasing decisions upon what I say or do. *I want you to base your decisions upon what the science tells you.* Science isn't interested in your credit card number. Learn to use knowledge as a tool to guide you instead of just blindly following the edicts of a media sage. The so-called sages of the audio media were not sent from the heavens on a divine mission to purify the marketplace and shield the helpless from charlatans. Everyone in this business is selling something, including the sages.

What then is the consumer's best choice of action? Become your own expert. I would be overjoyed to learn that many of you, after reading this book, went to the library to learn more about electronics and acoustics. Sometimes being a nerd is a good idea. You don't have to tell anyone that you read a book on electronics. It will be a secret between you and me. If you can absorb the basic concepts set forth in these few pages, then there should be no stopping you from continuing on. Most of the math you will encounter is no more difficult than simple algebraic relationships requiring only addition, subtraction, multiplication, and division.

How do you determine if your system sounds right? Easy! Attend a few live, musical performances. Whatever venue you desire, there are always opportunities to experience live music. The best types to learn from are nonamplified acoustical instruments and vocals, which are predominately orchestral music. Virtually all pop music (including jazz, folk, rock, and blues) is heavily amplified, mixed and equalized and then pumped through another, often substandard, sound system. So how do you know if your system produces pop music accurately or not? You don't. Don't worry, you aren't alone. Since there is no pure acoustic source, no baseline exists with which to make a comparison. Either the system produces an acoustical experience that is pleasing and satisfying to the listener or it doesn't. The only person that can make that determination is the individual using the system, and that includes you, me, and everyone else.

Now it's my turn to prove what I have done. The next section contains the actual designs of my products. Complete working schematics, parts lists, and checkout procedures are included. Anyone with the necessary skills to assemble electronics can build my equipment and find out for themselves if I am telling the truth or am just another carnival act. Anyone can say anything. Proving your claims is another matter. Well, here's the proof.

PART TWO
PROJECTS

*"If they don't understand it,
it doesn't exist."*
—Bruce Rozenblit

WHAT DO YOU NEED TO KNOW?

I would never claim that *anyone* can build these projects. Obviously, the more experience one has with electronics the better. The projects represent varying degrees of difficulty. The novice should start with a simpler project before attempting a more difficult one.

The circuit explanations assume a basic elementary knowledge of electronics. If you're not quite there yet, you should read my first book, *The Beginners Guide to Tube Audio Design*. This book is a straightforward distillation of electronics engineering principals that teaches lay people how to design their own circuits. If your needs are even more fundamental, then go to the library or book store and pick up a title that covers introductory electricity and electronics. A good test of competence is to check and see if the author holds at least a Bachelors Degree in Electrical Engineering. If the author has been to school, the odds of getting credible information are greatly improved. There are also many wonderful texts from decades past that are available as reprints (I have many in my library) but most are a little too difficult to digest if you are just starting out.

Some projects require extensive voltage measurements during check out. You don't just slap them together and throw the power switch on. These measurements involve working around dangerous voltages. You need to know how to use a digital voltmeter (around $30) and interpret the readings. This is very straightforward and simple, but it is a skill that some do not have, nor wish to acquire.

It would be wonderful if you owned an oscilloscope and signal generator. Scopes are available now for less than $400 (dual trace, 20 MHZ, quite powerful) and you can get 5 MHZ, single channel units for slightly more than $200 that would be perfectly adequate for check out and trouble shooting. They are your eyes into the workings of an electronic circuit. A simple signal generator can be had for around $60, and first class units are less than $200. Experience is your best teacher when using test equipment and there are many books commonly available that can help you. Part of my plan for this book is turning unsuspecting audiophiles into proficient electronic hobbyists. Hopefully, if you understand how the equipment works, you will develop an appreciation for superior design and demand it for your dollar.

You will have to know how to use a soldering iron and work with wire. Again these skills are easy to acquire, and absolutely essential for completing a project.

Cutting holes in the chassis is not that difficult. This work requires using drills, chassis punches, files and a nibbling tool. If you don't know which end of a screwdriver fits in your hand, you are going to have problems machining a chassis.

The general procedure to follow in building a project is to first acquire all the major components and lay them out to ensure a large enough chassis has been selected. Then buy the chassis, mark all the holes and cut them out. If desired, now is the time to paint. Mount all of the main components where any connections will be made. Wire it up and then check it out.

MACHINING A CHASSIS

Working with sheet metal takes a little getting used to. Unlike wood, sheet metal can bend or even tear if you don't use the proper tools and techniques. The operations needed will be cutting small round holes, cutting large round holes, and cutting rectangular holes.

It is far easier to work with aluminum than steel because it is much softer. Steel is hard and leaves sharp little slivers that can easily cut your fingers. Since I don't enjoy bleeding, I always use an aluminum chassis.

For round holes of diameter 1/4 inch and smaller, a sharp drill bit works well. For holes larger than this, the drill bit usually chatters, jams and severely distorts the metal. One way to minimize this is to use several drills and increase the hole size a little at a time. Increments of 1/16 inch are in order. The larger bits always chatter when they break through and you can never get a clean hole. For good hole center accuracy, always drill a pilot hole of no bigger than 1/8 inch. Unless you are using a drill press with the work piece clamped, the larger the bit, the greater the tendency for the bit to drift off center. It's not a question of when but of how much.

Don't use a high power drill and be sure to keep the RPMs down. Slower speeds, in the range of 500 to 750 RPM, cut more cleanly and don't load up the bit. A powerful drill is dangerous because if the bit jams (and it will!), the drill will spin out of your hand. That really hurts because your hand is gripping the drill tightly and since it is attached to your arm, it has a hard time spinning at 1200 RPM. A small cordless model is best.

For larger holes, always use a chassis punch. They are a little pricey but will last forever. Commonly needed punches are 1/2, 3/4, and 1-1/8 inches in diameter. The holes they produce are always clean with no ragged edges. Apply a couple of drops of machine oil to the punch to help it cut more easily.

Draw lines on sheet metal with a scribe which scratches an accurate line in the metal. Use the scribe to make a small indentation for the hole centers. Then use an automatic center punch to enlarge the indentation prior to drilling. An automatic punch is much more accurate then the kind you hit with a hammer. It makes quite a bang when the spring-loaded mechanism drives the punch so I often wear hearing protection. The metal chassis acts like a diaphragm and really amplifies the sound.

Whenever you drill through, burrs will be left on the other side that must be removed. I have found that a ½-inch drill bit held in the hand, is quite effective at removing the burrs. A standard de-burring tool's cutting angle is too steep. You might need to use a rag to hold the drill bit as the side flutes can be sharp.

Components, like tube sockets which require two or more mounting holes with tight precision, should be handled as follows. Always make the major hole that accommodates the body of the device first. Then, place the device in the hole and mark the location of screw holes for the mounting flanges. This compensates for most of the error in making the big hole. If you drill all three at once, they will never come out right. I think this has something to do with Demons or evil spirits.

Sometimes holes still end up a little off center. Needle files are then used as "hole stretchers." Place the component on the chassis and mark how much to file with the scribe. A few quick strokes of the file and the problem is solved. No one will ever know.

Cutting rectangular holes is more difficult. The most important rule to follow is to take your time. Mark the shape of the opening on the chassis. Take a 1/4 inch drill and remove most of the material by drilling successive holes that just overlap. Now use the nibbling tool to square up the hole. The cutting head will probably be on the inside of the chassis and your hand on the outside. This makes it hard to see where the cut will be made. Sometimes it is helpful to draw the hole on the inside of the chassis to alleviate this problem. Otherwise, nibble as close to the line as you can without going over and then finish the job with a ½-inch wide flat file. The metal cuts quickly so use little pressure on the file. Most devices, like power cord connectors, have overlapping flanges that cover the hole so perfection is not required. Now place the device in the hole and mark and drill for the mounting screws.

If you wish to paint the chassis, the best time to do it is after all the holes are cut. I use 150 grit aluminum oxide sandpaper in an orbital finishing sander and give the chassis a thorough once-over. This removes all scribe marks and roughs up the metal to accept the paint. Wash with an abrasive kitchen cleanser or Brillo pad to remove all the metal dust and manufacturing oil. Use a top quality, aerosol appliance-grade paint and primer of your choice.

Now here's the bad news. The paint instructions will tell you that it dries overnight. That is true, but it doesn't cure overnight. That means it is still soft and scratches easily. It could take three or four weeks for the paint to fully harden. I have never been able to wait that long. After a couple of days, it's full speed ahead and I start building. I know the paint will be scratched somewhere, but how long can you just sit there and watch the thing dry?

MOUNTING THE COMPONENTS

The next step in assembly is bolting all of the components to the chassis. Tube sockets and terminal strips require size 4/40 by 1/4 inch long machine screws. Low profile "binder head" screws work best for most applications. Sometimes "flat head" screws are required as the surface of the head must be nearly flush with the chassis for clearance with the tube. This will require countersinking the screw head. Small transformers are usually held with size 6/32 by 3/8" screws, and larger ones with size 8/32 hardware.

A set of screwdrivers with blades of 1/8, 3/16, and 1/4 inch are essential. A set of nut drivers are equally indispensable. Always use an internal tooth lock washer under each nut as it prevents loosening over time and eases assembly by grabbing the nut as the screw is tightened so it doesn't spin. For tight locations, a screw-holding screwdriver is a real time saver. If you don't have one, you can always tape the screw to the screwdriver with a piece of masking tape to get it started.

Don't over tighten small screws as you can easily shear them off. This isn't about winning the strongman's competition.

Mounting components on the front and rear of the chassis first is always easier. Then mount the components on the top. If you don't, you will be constantly scraping your knuckles on top-mounted components as you hold the nuts for the front and rear-mounted items. Make things easier for yourself and mount the transformers last.

WIRING

If there is one area in audio loaded with misinformation, it's wiring. Once you understand the magnitude and breadth of sonic changes that can be implemented by altering circuitry, it becomes a very difficult proposition to believe the many wondrous claims made for replacing a 3-inch piece of wire.

I select wire the old-fashioned way by matching the insulation to the application. Tube amps are hot places, so wire with high temperature insulation is a benefit. I always use 105 degree C., U.L. rated 1015, PVC insulated hook-up wire. It has a thick, tough insulation that can take a lot of abuse which, believe me, it will receive. The high temperature allows it to withstand the heat of soldering very well. The voltage rating is only 600 volts, but that is a U.L. standard rating. The insulation is thicker than other types of PVC insulated wire rated for 1000 volts. So for our purposes, 1015 wire will work at any conceivable voltage which we will encounter.

The conductors are tin-plated copper, which is the standard for all electronics assembly. The tin plating makes the wire much easier to solder. At this point I don't want to delve into the myths about secret alloys, plating compounds, geometries, crystalline structures and other topics that permeate contemporary audio literature and culture. They are far too numerous to mention and the previous section clarified many of these issues.

There are many other types of hookup

wire that will work just as well. Whatever you use, just make sure it has at least a 600 volt and 90-degree C rating. Wire with an 80-degree rating will work but it doesn't hold up as well to soldering. It is perfectly functional but it can't take as much abuse. Hookup wire is not expensive. If you are paying more than 15 cents a foot, look for something else. Buy it in 100-foot spools for best economy. You'll be surprised how fast you can use up a spool of wire.

For most low current wiring up to a couple of amps, 22-gauge is adequate. For current levels up to about 5 amps, 20-gauge should be used. Any application that requires currents over 5 amps should be wired with 18-gauge wire. These are rough break points; there's lots of room for overlap. Another consideration for wire size is how much physical abuse the wire will be exposed to. If it will be tugged and pulled, go with 18-gauge for the superior physical strength. I don't use wire smaller than 22-gauge because it is too easy to break.

A common concern among audiophiles is the choice between solid or stranded hookup wire. I've used both and can't tell any difference in sound quality. Solid conductor is easier to work with because you don't have to deal with all those little strands not threading through the terminal. Personally, I prefer to use 20 gauge solid conductor wire for rapid assembly. It is thick enough to give it excellent mechanical strength. I don't recommend using 22-gauge or smaller solid conductor wire as it breaks too easily. Any applications where flexing will occur should be wired with stranded wire.

Connections are made at terminal strips. Many sizes and configurations are available. They are positioned as needed in the chassis. Often they have one terminal that is grounded to the chassis. For most applications, that terminal should not be used. For minimum noise, the circuit should be connected to the chassis at only one point. Otherwise, ground currents could flow through the chassis and create noise.

To minimize the possibility of circulating ground currents in audio cables, I often make the chassis ground connection to the circuit through a small 10 ohm resistor. The best place to connect the resistor is at the input jack. Do not confuse this with the circuit ground. The circuit ground should be made at one point at the power supply to minimize ground loop hum. The input jack is directly connected to the circuit ground. The ground terminal of the input jack is then connected to the chassis through the resistor. This of course requires the use of an insulated input jack.

It isn't absolutely necessary to use the 10 ohm resistor with the insulated input jack. I have built many projects with the jack directly connected to the chassis. It is just that the diversity of installations in the marketplace varies considerably. As a manufacturer, I have to compensate for the worst possible conditions.

The ground wire of the power cord is always solidly connected to the metal chassis, which is necessary for electrical safety.

There are solders available that are billed as "no clean," meaning they leave virtually no flux residue. This makes for a neat and tidy job. Kester 245 is one type, but there are others that are just as good. Conventional rosin core solder will also give excellent results. Solder for electronics is 60/40 tin/lead respectively. The 60/40 means 60% tin, 40% lead which has been the standard for electronics use for many decades. Expensive alloys and exotic varieties are unnecessary. Lead is toxic so don't forget to wash your hands after soldering. Although it's safe to work with, you don't want to eat any of it.

The fumes produced by soldering are irritating to some people. In any event, it's not the best substance in the world to breathe. When doing a lot of soldering, I often turn on a small fan at the end of my work bench to blow the fumes away. For serious fume control, you can get bench top devices that draw the fumes away and trap them in a filter. They are somewhat expensive at about $100, so I still use my fan.

Soldering irons are inexpensive and commonly available. You don't need to go farther than your local Radio Shack to find a decent iron. A 25 watt unit will work for most applications and a 35 watt iron will be needed for large, heavy connections. Quality irons can be purchased for $15 to $20 and premium units for $35 to $40. Avoid anything less than $10. I have found that a chisel tip or screwdriver tip works better than a conical tip for tube work. If you want to go first class, get a temperature-controlled iron. They do a super job but will set you back more than $100. Temperature-controlled irons let you select the tip temperature for the application, greatly eliminating the need for more than one iron.

In general, when making a solder joint, the less solder used the better. Just cover the wires and terminal with solder, don't load up a big blob on the terminal. A finished joint will be smooth and shiny. Solder has to be motionless for the few seconds it takes to solidify. If you use too much, it will try to flow under its own weight and cause a "cold" joint. These are easy to find as they are dull and whitish. Wires should always be mechanically fixed to the terminal by bending and crimping so they don't move when soldering, assuring a good joint will be made.

One of the most common pitfalls, when doing this type of soldering is missing a wire when there are several on a terminal. What usually happens is that one wire is attached at the bottom of the terminal and the others are wrapped around the top. You solder the top wires and the lower one gets missed. Now you have an intermittent connection that can drive you nuts. Always check each wire to make sure this does not occur. Give any questionable wires a little tug. If they move at all, then re-solder. Take your time, don't be in a hurry.

Whenever a joint must be re-done or wires moved, the solder must be removed. Pump action vacuum devices do the job quite well and are less than $10 with premium units at $15 to $20. Sometimes there may not be room to use the vacuum device. In that case, solder braid can be used to clean off the joint. Just place the braid on the joint, apply the iron, and solder will be wicked into the braid.

Another common failure is broken wires. If one is careless when stripping insulation, the conductor could be nicked. Repeated bending during installation can cause the wire to break at the nick. You might not see the break because it usually occurs just under the insulation and everything looks fine! Again, the culprit is found by tugging on wires. This problem is confined to solid conductor wires. Stranded wire can also cause problems as some of the strands can break, causing more stress on the remaining strands, which eventually could fail. Avoid a lot of trouble and re-strip any nicked conductors. The use of an inexpensive stripper (less than $10) that has preset dies for each wire size will sharply reduce these problems. I wouldn't do this kind of work without one.

Routing of wires is a skill. Professional wiring assemblers can do an absolutely beautiful job at lightning speed. I have yet to master this art form. Remember, never wire point-to-point in a straight line. Always wire in a grid pattern, bending 90 degrees as required. When you are done, all of the wires should be essentially arranged in only two directions. That is, up and down and left and right. They can then be bundled together with nylon wire ties to give a quality job. Use enough wire ties to keep the cables from bouncing around. Always leave some slack so no wires are under tension.

There are some wires that should never be bundled together. Any wires carrying AC power should be segregated from wires carrying audio signals. This includes AC filament power and the power line. You will find that it is physically impossible to totally segregate filament power from the audio. These signals often will cross paths and be in close proximity to each other. Don't worry about it. Problems only arise if you bundle these wires together for long distances and a few inches in close proximity isn't going to make any difference.

Traditionally, AC power cables are twisted together to minimize radiated fields from contaminating the signal. In power amps, this is a good practice. It is not the end of the world, however, if you don't. It's just not that critical. In preamps, twisting is definitely necessary but most modern preamp designs use DC for the filaments thereby eliminating the interference problem. Distance is quite effective at reducing interference from radiated fields. I always mount the power transformer, power cord, and switch on one side of the chassis and the audio input on the other. Also, it is always preferable to mount the active device of the first stage near the input jack to reduce the possibility of picking up any hum and noise. The larger the amplitude of the audio signal, the more resistant it is to noise pick up. This is because the ratio of the signal to the noise is greater, thereby masking the noise, rendering it less audible if not inaudible. The sooner the signal is amplified, the greater the masking effect. Again, this is not super critical. Just don't route input signals haphazardly all over the chassis.

Make sure the wires are not in physical contact with anything that gets really hot, like a big power resistor. It is also important that the wires are not pressing or chafing against any sharp surfaces that could cut through the insulation and cause a short. A little care and common sense goes a long way in doing a good wiring job.

The more comfortable you are, the better job you will do and the more enjoyable the work becomes. Don't build projects on the floor. Set up a small work bench or table with a good light that can be dedicated to the project. This kind of work doesn't require a lot of space but some small amount of real estate must be allocated or mistakes and much frustration will follow. Grab the wrong end of a soldering iron once and you will heed my advice thereafter.

The projects in this book have varying degrees of difficulty and the reader needs to be cognizant of his or her skill level before attempting one. If I had to rank them from easiest to build to most difficult, the list would be as follows:
1. Grounded Grid Preamp
2. Phono Preamp
3. 150 Watt Compact Amp
4. Single-Ended With Slam
5. The Transcendent OTL

They all have a distinctive sound. What would be the point of having four different amplifiers if they all sounded the same? Many amplifier companies try to make all of their products produce the same sound and use their advertising clout to "prove" that sound is the best. My approach is completely different because I am covering all of the basic configurations and pushing the performance of each type to levels *higher* than the standard products that grace most dealers' showrooms.

RULES OF THE GAME

1. Please do not contact the author with questions about changes, modifications, and alternate parts selection. Build the projects exactly as shown. If you wish to alter the circuitry, then do so at your own risk. If your idea doesn't work, you can always change it back to the original design.

2. If there is something that you don't understand or are unable to interpret, then you may contact the author with email at:
tubehifi@worldnet.att.net
The author appreciates receiving email praising his work. If you have nasty things to say, don't bother to write.

3. All voltages shown on the schematics are approximate. Expect variations of 5 to 10 percent.

4. All resistors are 1/2 watt, 1% metal film unless shown otherwise.
All 2 and 3 watt resistors are 5% metal oxide.
All 5 watt resistors are 5% wirewound.

5. All capacitors larger than 1.0 mfd are electrolytic.

6. Capacitors with higher than specified voltage ratings may be substituted.

SAFETY PRECAUTIONS

1. Always wear safety glasses when cutting metal, cutting wire, or soldering.

2. Never place both hands inside an energized chassis. Never let any part of your body come into contact with an energized chassis.

3. Always unplug a unit from the wall outlet and discharge all capacitors before making circuit modifications.

4. Never hold wires or parts with your mouth.

THE TRANSCENDENT OTL

This is it! It doesn't get any better than this. If you want clarity, detail, resolution and accuracy, nothing beats an OTL. No transformer-coupled or solid-state amplifier can match it, and this is the best OTL there is. A transformerless tube output produces a sound that cannot be duplicated with any other technology. It is light, airy, open, tremendously fast, yet totally musical and natural without any coloration of any kind. Two power levels are presented in monoblock format at 25 watts and 80 watts. The higher-powered unit can drive all but the most difficult loads. It can make a solid 50 watts into 4 ohms. That is plenty of power for most speakers. Output impedance is a very low 0.4 ohm. This is no weakling. I would pit the 80-watt unit up against any other amp at any price. Of course, it can't play louder than a 500-watt monster amp, but the issue I raise is quality of reproduction.

I would characterize the difference in sound quality between the two as the 80-watt model being slightly more relaxed in the mid-range and treble, which is a subtle but noticeable improvement. Otherwise, the tonality is the same. The biggest difference between the two is in the bottom end. When it comes to bass drive, there is no substitute for horsepower. The 80-watt amp has much more slam and impact than the 25-watt unit. I would say that the bass is no deeper, but it can generate a lot more of it. That is the most significant benefit that an increase in power brings for any amplifier.

The net heat rejection for these amplifiers is less than half of any other OTL in the world. The 25-watt amp will only put 300 watts of heat into your room per pair. That is low enough that you can operate the unit for hours without any appreciable temperature rise in most listening rooms. Don't let the low power rating fool you. It will drive a lot of speakers. The 80-watt model will take up more space and produce twice as much heat, at approximately 600 watts per pair, which is definitely in the noticeable range. This may sound like a lot of heat, but competitive OTL designs have heat rejections of 1500 to 1800 watts which are impossible to live with unless you keep your amp on the front porch. In fact, any large solid state class A amp will put out at least this much heat–if not 800 to 1000 watts. So the amps could definitely be considered reasonable with respect to heat rejections compared to other products that are currently available.

The driver circuitry is exactly the same in both versions. The only difference is that the 80-watt amp uses eight output tubes instead of four and has a huskier power supply for the extra outputs and filaments. If desired, you could take its power supply and drive two 25-watt amps, all on one chassis. That's what I do in the commercial stereo version. However, it is easier to build the monoblock configuration since there is much less wiring congestion.

Monoblocks also give greater performance with regard to peak power and lack of crosstalk.

As good as these units are, they are extremely hard to sell. Because of the many failed designs that have been put on the market over the last 30 years, most audiophiles are afraid of an OTL and won't touch one. The audio community's conception of an OTL is based upon the many disappointments and unacceptable products that have resulted from substandard engineering or, I am sorry to say, no engineering at all. People have been working on developing OTL amps for decades, never once solving all the chronic problems of extreme heat, terrible reliability, weak bass, and incompatibility with dynamic speakers. Audiophiles have asked me many times: "How could you have possibly come up with a better design when all those many people from all over the world have been working on this for more than 40 years?" The answer is an easy one. The wrong people have been working on the problem.

Tube amplification equipment is very forgiving and it is easy to make it sound good. Consequently, the untrained or novice, provided he can secure enough capital, can go into the tube audio business. The products may have good sound, but that doesn't mean they won't break down twice a week and three times on Sunday. It would be interesting to see the actual failure rates of much of the highly regarded tube equipment. It is no wonder that tube equipment has the reputation of being unreliable. It's not the tubes' fault. When properly constructed, tube gear can easily go for 20 or even 30 years without failing. It just has to be put together properly. That requires schooling, training and experience.

This amplifier is dramatically different from anything that has ever been produced. How do I know that? I have been awarded a United States patent for the design. If anyone in the world had ever filed a patent for, or published a design similar to mine, I could not have received the award. This is not my opinion. It's a fact.

The problems of the OTL are all practical in nature. They were just impossible to live with and only people willing to put up with the most esoteric of an esoteric field, ever bought them.

What I have created is what I call a "user friendly" OTL. You just plug it in and use it, no fiddling required. When you want music, you turn it on and it works, day after day. The amps will easily drive conventional cone speakers and are not restricted to electrostatics in any way. All OTLs benefit from higher load impedance but this is not an absolute requirement. Any 4 ohm speaker with a sensitivity of 92 dB or greater is a suitable candidate for the 25-watt model as the amp will produce a clean 15 watts into 4 ohms. Any 8-ohm speaker with a sensitivity of 89 dB or greater is also a good match. For some reason, a single-ended amp that can produce only 5 watts is considered adequate but a push-pull amp that makes 25 watts is considered underpowered. This is another of those mass marketing mysteries that I don't understand. A clean 25 watts is really a lot of power.

A proliferation of speakers are available that are intended for use with low power amps. They can produce quite high sound pressure levels on 15 to 25 watts. Speakers with impedances that fall into the 2 ohm range are unsuitable. The 80-watt version, with its much higher current capability, can tolerate low impedances much easier and drive all but the toughest loads. Any 4-ohm speaker with a sensitivity of 88 dB or greater or 8-ohm speaker with a 86 dB rating will work well–and that covers most of them. My point is that these amps do not require a 16-ohm speaker and will drive the majority available. The myth that they cannot is a result of the other OTL products that cannot drive a 4-ohm load.

Driving low impedance speakers requires two things: sufficient current and a low output impedance. These are "must have" conditions that cannot be compromised. This is an engineering fact.

Audiophiles often get confused when distinguishing between a published *fact* and a published *opinion*. Anytime I say something is a fact, that means it can be measured and verified with instruments in a repeatable fashion, just like a scientific experiment. Water always flows downhill. That is an engineering fact. A person can spend an entire lifetime trying to get water to flow uphill but it never will. Due to the proliferation of unqualified people in the audio industry, there are many more opinions than facts in circulation which, unfortunately, costs consumers a lot of money. I suspect the mass marketers have something to do with this.

The only practical way to meet the requirements for driving 4-ohm speakers with an OTL is to use enough of the proper tubes and apply negative feedback around the output stage. If you elect not to use feedback, so many tubes are necessary that a forklift will be required to move the amp and an exhaust fan will have to be installed to pull the heat out. Units that claim no use of feedback have a much higher output impedance and consequently much poorer bass response.

A loudspeaker is a type of motor with a linear armature. It goes in and out instead of round and round. It has a mechanical compliance similar to a car's suspension system that keeps the wheels from bouncing wildly when you hit a bump in the road. When the car's spring is compressed, energy is stored which the shock absorber dissipates in a controlled fashion, thereby *damping* the motion and preventing an oscillation. Otherwise the car will bounce up and down uncontrollably. The mechanical suspension of a loudspeaker operates in a similar fashion. The compliance of the loudspeaker's suspension is reflected into electrical impedances that the amplifier must drive. These impedances are reactive in that they store energy just as the car's spring does. The stored energy must be dissipated somehow. The place it goes is across the impedance of the amplifier's output stage. So the lower the output impedance the greater the ability for the amp to dissipate the stored energy and the better control the amp has over the cone motion. This results in more accurate and faithful sound reproduction. The ratio between the amplifier's output impedance and the speaker's impedance is called the *damping factor*.

I have found that a damping factor of 10 gives good control and 20 is excellent. Damping factors greater than 20 don't seem to yield much further improvement. Amplifiers with damping factors less than 5 cannot possibly extract the best sound out of the loudspeaker. Audiophiles have told me of outstanding bass response obtained with zero feedback amplifiers that have low damping factors. What they are actually hearing is a woofer out of control generating more bass energy than was in the original program material. This is particularly evident during sustained tones. Listen to sharp percussive bass, and the sound falls apart and gets mushy. In fact, the impact of the bass is rounded off so much that it almost sounds like a continuous tone. The woofer simply can't start and stop quickly enough. All of my amplifiers have a low output impedance, giving the bass a quick, tight and precise sound. You hear what the music is and the speaker works to its maximum ability, not like a boom box generating sounds that were never there in the first place.

To achieve high reliability, tube longevity, and low heat in an OTL, it is absolutely essential to use as low a bias current in the output stage as possible. Class A, or anything approaching class A, is asking for trouble. The only way to achieve any substantial power output is to overdrive the tubes to their maximum limit. Here is where the trouble starts.

The more current a tube has to produce, the hotter it operates and the faster it wears out. Music is dynamic in nature. Even highly compressed pop music has a peak to average dynamic range of 20 db. That means if the amp has to produce 100 watts during peaks, it only needs to produce 1 watt for the average signal. Low-level signals are then small fractions of a watt. Obviously, the percentage of time an amp operates at maximum power is very small indeed. So peak power capabilities more closely correlate to how an amplifier actually performs in a real-world application.

Most OTLs produced by other firms (many are out of business) run the output tubes at maximum plate dissipation. This is a big mistake. During peaks, the tubes go into severe overload. Since the plates are already at maximum, there is no capacity to absorb the thermal shock of the short term overload. The result is that the tube gets very hot and is prone to fail. Your listening room gets really hot, too.

If the output tubes are biased at low currents, the resultant plate temperatures are much lower than the allowable maximum. When the peaks come along, plate temperatures do go up, but the *average* dissipation is still well within a tolerable range. I call this a "thermal reserve." Hence, tube temperature and reliability are primarily determined by the bias current setting. Class AB operation is therefore absolutely essential for reliable OTL operation.

We're just knocking audio myths off the table with each paragraph. Another is about to fall.

How do we deal with all of that awful crossover notch distortion produced by class AB operation? The answer is another question: What crossover notch distortion? My friends, I can't find any. I've been looking for years and nary a notch can be found. I can't find them in my OTLs or in my work with transformer coupled amps. Honestly, I don't have the vaguest idea what the rest of the world is talking about when they say class A push-pull tube amps are more linear because they don't have crossover notch distortion. The only conclusion I can draw from experience is that there is absolutely no reason to operate any push-pull tube amplifier in any mode other than class AB. My best guess as to why this myth exists is that it is just another mass marketing ploy to get people to get rid of their economical, long-lived class AB amps and buy very expensive, failure prone class A units. If the United States during World War II produced tube equipment like some contemporary high end amplifiers, we would all be speaking different languages.

In addition to the bias setting, the configuration of the output stage is crucial to making the amp run as cool as possible. The series connected (or totem pole) push-pull pair is absolutely the most thermally efficient possible. That's a fact. The positive half cycle is provided by the upper tubes and the negative half cycle is provided by the lower. By virtue of the connection and class AB operation, the tubes are off roughly 50% of the time during moderate to high current operation. This gives them a chance to cool down, greatly relieving thermal stress. Designs that incorporate a floating or differential output require all tubes to conduct at the same time, thereby doubling plate heat rejection. Better open a window.

Because the output tubes are overdriven during high current operation, they become rather nonlinear. If we didn't overdrive, power output would be limited to a few watts, which is insufficient and hardly cost effective. Feedback is necessary to linearize the output stage under these conditions. Designs that boast of no feedback must have a lot of distortion at high power levels because there is no way to avoid tube nonlinearity in the overdrive condition. I can see it coming. Somebody is going to read this and develop a zero feedback OTL with a dozen tubes that puts out only 15 watts. I just hope they include a range hood with the shipment. Of course, then the mass marketers will get involved and promote an "audiophile grade" range hood that retails for only $5000– installation extra.

The heat problem affects not only the reliability of the output stage but the entire assembly as well. All of the other components in the unit get baked by the output stage. Electronic components are prone to fail if they get too hot. So by cooling off the output stage, the rest of the amp is given a heat reprieve thereby increasing reliability.

Another factor necessary for reducing output impedance, greatly improving bass response, and making terrific sound, is a direct coupled connection to the loudspeaker. Output capacitors are expensive, failure prone, and greatly restrict bass response. Solid state amps eliminated output capacitors decades ago for the same reasons. This mandates the use of complimentary power supplies (both positive and negative) for the output stage. This and the preceding paragraphs illustrate why the output stage for my OTL is configured as it is. The selected configuration has the lowest heat rejection and highest reliability and performance of all possible choices. That's a fact.

As you can see, I'm setting the practical requirements as the top design criteria. Instead of playing around with circuits to see which sounds best, my initial task is to get the thing to work properly, maximize reliability, and make it easy to live with. I know I can make it sound good later. That will just take some more engineering to work out. My pathway of development is probably exactly the opposite of most of the rest of the industry. If more companies followed my procedure, tube equipment would have a much better reputation.

The next myth that must be smashed is the one that OTLs blow up speakers. This one does not. A little more engineering must be mixed in with the brew.

Before a protection scheme can be designed, all possible failure modes must be examined. As a manufacturer, I have had the opportunity to test hundreds of tubes, revealing the problems that can occur. I can find most bad tubes upon initial checkout. Sometimes, however, an output tube will fail in the first few weeks of use. In electronics terminology, this is called an "infant mortality."

The tube I use in the OTL is the best I have found for the application, the EL-509. The big tetrode has many advantages over other possible candidates. Specifically, it is much more consistent, in good supply, rugged, and sounds a heck of a lot better than anything else I've tried. The tube is used in a triode mode by connecting the screen grid to the plate through a 100-ohm resistor. The plate impedance for this configuration is less than 150 ohms, which is very low indeed. There are beam plates (pins 2 and 7) which are left open. In tetrode mode, they would be connected to the cathode. Experiments in tetrode mode did not result in any increase in power. An OTL with a full tetrode or pentode output stage would be dramatically more complicated as floating regulators would be required for the screens.

Three possible failure modes have been identified, two of which are rare and one which has never happened as far as I know. One failure is a short between the cathode and the filament. When this happens, the tube doesn't light up. In one customer's amp, this failure caused the power supply fuse to blow. The tube was replaced and there was no damage to anything. The few other times this has occurred, no fuse blew and again no damage occurred.

The failure that everyone is most afraid of is the one that has never happened in the field: the plate to cathode short. This has the potential of strapping the power supply directly to the speaker. I hate to give anyone nightmares, but all solid-state amps have the same problem and they *do* blow up speakers when output transistors short. The energy stored in a big high-power, solid-state amp can do a great deal of damage in an instant. So how do I know the protection works if the fault never happens?

Before discovering the 509, I had several years experience with the 6AS7G. This tube experienced arcing faults from plate to cathode quite frequently. When the arc occurred, a tremendous pop could be heard. The energy of the power supply would be dumped into the tube, always ruining it. At that stage of development, I did not include any resistance in the cathode circuit (R31-R34, R45-R48). The magnitude of the fault was enormous, probably more than 1000 amps. The power supply rail fuse would blow virtually instantaneously. I would estimate fuse clearing time in less than a millisecond. No speaker damage ever occurred. Why?

Several circumstances protected the speaker. Most of the fault energy went into the tube. The brunt of the energy was absorbed by the tube which always sustained physical damage. Although the magnitude of the fault was extremely high, it was over within an instant. No steady state DC current flow was allowed to become established. There simply wasn't enough time for any voice coil heating to occur. The fault was a spike, an AC condition whose energy was contained in very high frequencies. The inductance of the speaker and associated cables opposes the energy of the spike. No DC can flow until this reactance has charged. Before that could happen, the fuse would blow, thereby shutting off the current. Also notice C11 and R40 strapped across the output. Their primary function is for feedback stabilization but during a fault they act like a snubber circuit to shunt any high frequency fault current into ground. So, because of load inductance and the snubber circuit, very little of the massive fault ever went into the speaker.

Looking at the tube specs sheds some light as to why the 6AS7G would often fault and the EL509 does not. The 6AS7G has a plate voltage rating of 250 volts. The amp applies 170 volts across the tubes. The 509 has a plate voltage rating of 900 volts with a pulse rating of 8000 volts! No wonder it doesn't arc over. The physical structure of the tube allows it to withstand the much higher electric fields associated with OTL operation which are particularly exacerbated by the overdrive condition.

Since those early days, I have improved the fault protection considerably. There is now a 2 ohm resistor in the cathode of each output tube. The small resistance reduces the peak fault from an estimated 1000 amps to a maximum of 75 amps. Fault energy is proportional to the square of the current, so the fault energy is then reduced by a factor of 200 times. If a tube ever does arc over, the power supply fuse would blow in a couple of milliseconds and the tube would not be damaged.

The cathode resistors provide an additional benefit. They create just enough degenerative current feedback to equalize the current through each output tube during high-power operation. This helps prevent any one tube from hogging all the current, becoming overstressed, and failing. The cathode resistors increase the output impedance by about 10% and consume a few watts of output power, but the benefits they create are well worth it.

There is one more fault that we have to protect against. That is the grid-to-cathode short. This failure does occur from time to time in the field. When the grid shorts, no bias voltage is available to modulate the current through the tube and it turns full on. The maximum current that a 509 can pass is just over one amp.

Each grid has a resistor connected to it in series. Its primary purpose is to prevent parasitic oscillations and a 1K unit will suffice for that task. I am using 100K resistors instead. When a grid does short, the large resistor effectively isolates each tube from each other by greatly limiting how much bias current can flow into the shorted grid. Then, only the shorted tube turns full on. If a 1K resistor is used, the bias voltage is pulled down for all tubes of that bank (4 in the 80 watt unit, 2 in the 25 watt unit) and they all turn full on.

The power imparted into the speaker from the fault is minimal. If the speaker has a DC resistance of 5 ohms, it will be subjected to about 6 watts of heating. If your speaker can't take that, you'd better get something that doesn't use a $5 woofer. The offending tube is easily found as it will be glowing red hot!

My protection scheme is just the opposite of what everyone else does. Typically, OTLs use a separate active circuit to monitor the output for DC and then disconnect the power supply or speaker or both. The protection circuit has to do three things: (1) recognize the fault; (2) make a decision about the fault; and (3) take corrective action. All that takes time. It takes 25 milliseconds to operate a relay *after* the circuit decides to do something. What happens if the protection circuit fails in any of these tasks? Damage to something most certainly will occur. A simple fuse will *always* work.

These amps do not have any separate protection circuits. In fact, they don't use any voltage monitoring at all! What I am doing is controlling the current by keeping it from reaching destructive levels for both the speaker and the amplifier. The protection is integral to the audio circuit, not an external network that has been added on. It can't fail because there is nothing there to break. The system is extremely reliable due to its utter simplicity. That, my friends, is the difference between good design and great design.

Now we get to the good stuff! The stage driving the outputs is the essence of my patent. Everything discussed so far are refinements and enhancements to the prior art. All of these are valuable, but not worthy of a patent award.

Only four U.S. patents have ever been issued for the series connected push-pull vacuum tube output stage. One to a Mr. Peterson in 1957, two to Mr. Futterman in 1956 and 1964, and one to yours truly in 1997. As you can see, these things are hard to get.

What I have discovered is terribly simple and so obvious that it's a wonder no one found it earlier. It seems that everyone else has been following in the shadow of Futterman. It must be my rebellious nature that drives me to try new things. Hero worship is not necessarily conducive to creativity.

There is a fundamental problem in driving this type of output stage. The lower tube has its cathode referenced to ground via the negative power supply. Any tube is driven by the audio signal voltage between its grid and cathode. Since the cathode is at ground potential, the driving voltage is also referenced to ground.

Most tube circuits work this way. The upper triode has its cathode connected to the output. This means that the cathode is not at ground potential but floating on top of the output signal. The output signal is in phase with the audio driving voltage at the upper tube's grid. This causes degenerative feedback for the upper tube. The lower tube does not experience any degenerative feedback by virtue of the cathode being grounded. The result is that the output stage is unbalanced. A means has to be developed to increase the drive signal to the upper tube to compensate for the loss of voltage gain caused by the degenerative feedback. (For a more detailed discussion with illustrations, see *My OTL Patent* at the end of the projects).

The traditional way to do this is to apply 100% feedback of the output signal to a phase splitter through a floating voltage regulator. This technique will amplify the upper drive signal by the amount needed to overcome the degeneration. But what about the lower? The amplitude of this signal is unaffected by the feedback because the phase splitter operates as a cathode follower for this output. Everything seems to be perfect. But there is a problem here.

The upper signal is increased in amplitude by positive feedback which means the gain through that stage is increased. The opposite happens for the lower signal. The feedback, while being positive, acts to amplify the upper signal; it is *negative* with respect to the lower signal and tries to reduce it. The lower signal's amplitude is not reduced because the cathode follower action constrains the amplitude to be about 95% of the drive into the splitter. The forward voltage gain through the system is increased for the upper drive signal and reduced for the lower. What happens then is that forward voltage gain *through the entire amplifier* is consumed to compensate for the negative feedback connection for the lower drive signal. The gain that is consumed is extracted from the previous voltage gain stages. The system is unbalanced because it has unequal forward voltage gains for the two split phase signals. This unbalance is the primary reason for the poor bass response when driving dynamic speakers.

When observing the circuit of the prior art on the oscilloscope, as the amplitude of the output is increased a proportional DC offset is generated that pulls the output negative. That phenomenon is what prompted me to start looking for problems. At that time, I had resolved to myself that OTLs just couldn't have a great bottom end.

When I finally realized what was going on, I tried the circuit that formed the basis of the patent and launched me on the path to becoming an audio manufacturer. The solution to the unbalanced gain problem was to simply compensate for the degenerative feedback for the upper triode and leave the signal for the lower triode alone! Implementing this requires an intervening stage of unity gain buffers where one is used to apply the corrective signal to the upper tube and the other buffer is used for symmetry and has no signal correction.

The immediate result of this discovery is dramatically improved bass response. At last, I had an OTL with a bottom end. Observing the circuit on the oscilloscope, the amount of negative DC offset is greatly reduced. There is still a slight vestige of negative voltage at maximum output, but it has been reduced to insignificant levels. As is, the circuit sounds so good and is so unbelievably simple to produce, I say leave well enough alone.

The upper buffer applies the corrective signal through a bootstrapped two-terminal shunt regulator. That's a lot of words for three zener diodes. A series pass three-terminal regulator will certainly work, but would be vastly more complicated. The zeners are extremely rugged and get the job done perfectly well. Resistor R16 feeds current into both the triode (V3, upper) and three zeners. The cathode resistor (R17 plus R18) and zeners are connected to the output. This connection forms a floating reference or a ground above system ground. The voltage at the plate of triode V3 is the addition of the DC setting of the zeners *plus* the audio output. The voltage at the end of the cathode resistor is the audio output. The difference between the two is just the DC voltage of the zeners. So the triode exists in its own isolated world, thinking it is sitting in an envelope of 450 volts, not realizing that it is bouncing up and down with the output signal. This is how bootstrapping works. The resultant audio signal output at the cathode of V3 upper is then the audio input at its grid added to audio output signal being fed into the speaker. The signal degeneration of the upper tubes, therefore, is completely compensated for.

Coupling capacitors are used to isolate the different DC voltage levels of the two buffers from the negative bias on the output tubes. Making the entire circuit DC coupled would be an accident waiting to happen. With no capacitive isolation, any DC voltage upset in the gain or buffer stages of the amplifier (like pulling out a tube) would cascade through the system with potentially catastrophic results. I have seen this in solid state amplifiers before where a 20-cent input transistor fails, which takes out the next stage, cascading all the way through to the outputs, burning them out and then the power supply! By using an AC coupling before the outputs, this cannot happen. In fact, I have had customers put the driver tubes in the wrong sockets and proceed to use the amp. They call me in a couple of days and question why the power seems to be low or there is a slight buzz. Switch the tubes and everything works fine, no damage at all. Some customers have used amps for months with a blown output tube and never knew it. I have even had people use the amps with a blown power supply fuse, which means only one bank of tubes is operating (all push and no pull), and not realize it. On each of these occasions, simply replacing a tube or fuse corrected the problem and no damage ever resulted to the circuitry. These events verify that my OTL is a very tough and rugged product.

The output stage and the buffer stage comprise two of the three sections of the amplifier. The third stage is the voltage gain section. The output stage has a voltage gain of 0.25 and the buffers have a gain of about 0.95 so all of the forward voltage gain has to come from somewhere else. This section needs to provide a gain of at least 500 times to overcome the losses of the other two sections and have enough gain left over to use for feedback. I am using a very conventional gain section that has been included in hundreds of designs because it works so well and is so very simple. The circuit is comprised of a high gain triode direct coupled to a common cathode phase splitter. By staggering the splitter's plate resistors as shown, the split phase signals are essentially equal in amplitude. Additional balance adjustments don't yield any performance enhancements.

The power supply voltage is set high because the stage must output linear signals with amplitudes of 200 volts peak to peak. There are many other configurations that will work, including solid state, but this gets the job done with the fewest parts. The first stage provides most of the gain at just over 70 times, while the second stage provides an additional gain of seven times. Everything is DC coupled up to the output stage, giving excellent low frequency stability and freedom from motorboating.

The open loop gain of the circuit is 42 dB. Feedback is set at 22 dB yielding a total closed loop gain of 20 dB. That is lower than most amps by 6 or 8 dB but still adequate for most installations. Many of my customers use passive preamps. An active preamp with 12 to 15 dB of gain is a good match. Use a preamp with too much gain and you will never be able get the volume control past the 9 o'clock position.

There are two compensation networks used. One is on the output comprised of R40 and C11. This type of network is commonly used in solid state amps to compensate for highly inductive loads. The other network is the feedback loop itself, simply R11, R12 and C4, all in parallel feeding R6. I have set the compensation to a slightly over-damped condition to enhance the amplifier's ability to drive highly capacitive loads. The unit can drive a 2 microfarad capacitor in parallel with an 8-ohm resistor while being fed with a 10 kHz square wave and not oscillate. That is a very tough test of stability. It is also stable with or without anything connected to the input or output.

The bias network is very simple. Instead of adjusting each tube, one bank is balanced against the other. The bias voltage for the upper positive tubes is set at minus 40 volts. The bias voltage for the lower negative tubes is then adjusted to give a zero DC offset voltage in the output, at approximately minus 210 volts. The zero is obtained by balancing the net current from one bank of tubes against the other. The bias is not regulated because the output tubes are very sensitive to plate voltage. Any variation in power line voltage will cause the plate voltage to change, resulting in a change of plate current. By not regulating the bias it is allowed to roughly track the changes in plate voltage and compensate for these aberrations. Regulating the plate voltage would be a reliability nightmare and would necessitate an outboard power supply chassis. There is a very good possibility that doing this may not improve the sound and may actually make it worse.

The power supply is conventional but crucial for proper operation. Many voltages are needed. The commercial model incorporates a custom-wound transformer but its function can be duplicated with commonly available off-the-shelf units.

Toroidal transformers are seldom used in tube equipment. I don't understand why not. They are half the size and weight of E.I. stacked lamination types, have better regulation, and most important of all, don't hum. Many highly regarded tube amps have a terrible mechanical hum emanating from their power transformers. This noise is every bit as objectionable as hum from the speakers but is often overlooked. Maybe the extra weight is a desirable feature. An audiophile once told me he was going to buy a single-ended amp because it weighed 90 pounds. He never heard it. I am serious. He wanted it because it weighed so much. Maybe I should offer an optional ½-inch thick lead bottom plate, if that's what it takes to make a sale. Of course it would be audiophile grade lead so I could charge another $1500 for it. I could say that it "had a pronounced effect in magnifying the upper midrange presence, particularly during periods of low humidity." I jest, but the sad side to this humor is that many people have spent a lot more than $1500 for products that do less than that lead plate would. At least you could use it for protection against nuclear radiation.

There is a minor downside to the toroid. If you close the power switch at the zero crossing of the power line voltage, you get a tremendous surge. Power switches don't like that. I have begun to use current limiting thermistors to nullify this problem. They have a high resistance when cold and the resistance drops to a fraction of an ohm as they warm up in about two minutes. Another important benefit is that they eliminate any large current inrush into the rectifiers and filaments when a unit is turned on. That inrush is very stressful and potentially damaging to electronic components. Careful listening tests revealed absolutely no change in the sound with the thermistor present. I consider it a $2 insurance policy.

Any large bridge rectifier contains a lot of capacitance. The capacitance will react with the inductance of the power transformer's secondary winding and create some nasty voltage spikes. These spikes have a lot of power and will permeate the entire circuit. They get into everything, even showing up on the circuit ground! They can be eliminated by installing capacitors to ground on the transformer secondary leads. The spikes are primarily high frequency energy and can be harmlessly bypassed to ground. The impedance of the transformer keeps the capacitors from being subjected to damaging current surges. These spikes are the cause of a lot of high frequency buzz in audio amplifiers. They have to be treated on the AC side of the power supply because the filter section won't stop them.

There is no DC connection of the filament supply to ground. This is to relieve voltage stress between the filaments and cathodes of the tubes. An AC connection is required to minimize any high frequency parasitic oscillations from these same elements. This is accomplished with a small capacitor that ties the filaments to ground. It is connected to the center tap of the filament for the 12AX7.

You don't have to use the transformers that I have specified. If you have something in the junk box that will work, by all means use it. If you don't know how to determine what else will work, then use the parts specified. The goal here is to feed the amp circuit with the correct voltages. The most critical one is the bias voltage. With the network shown, the desired result is minus 40 volts for the upper bank of tubes. The ends of the bias pot at are at minus 200 and 220 volts so center rotation of the wiper is near 210 volts.

The output stage should be plus/minus 170 volts, within a range of 165 to 175 under no load. Under full load, the plates will pull down to about 155 volts. Setting the filament transformer at 20 volts will produce about 25 volts DC under load. All filaments have to be run on DC to minimize hum. The high voltage transformer shown is a 200-volt unit wired as a voltage doubler. A 400-volt unit will work when wired with a full wave bridge. The desired output is between 550 and 575 volts at the first filter section under load. Regulating this voltage will cause no improvement in performance. That's really about it. Now it's time for assembly and check out.

CHECKOUT

The power supply can deliver a lot of energy, so a staged checkout procedure is necessary to prevent component damage in the event of a miss wire or a defective part. Follow this procedure and do not connect the amp into your system unless it has been thoroughly checked out. This isn't the type of product that can be dropped off of a conveyer belt into a box and shoved into the back of a truck. The following comments apply to all of the projects in this book.

The first thing to do is check and see if all of the power supply voltages are present. Do not install any tubes at this time. Install the power fuse, plug the unit in and turn it on. WATCH, LISTEN, AND SMELL.

If there is a short, the transformers will grunt and buzz loudly. Shut the unit down immediately. If you see or smell smoke, shut it down. Electrical components have a distinctive odor when they start overheating. Sooner or later, you will be introduced to this aroma. Start looking for a miss wire or possibly a diode or electrolytic capacitor that was installed backwards. That will act like a short.

If the short is massive, it will blow the line fuse. This protects the amp but makes it hard to troubleshoot. A close visual inspection is the next step. If you still can't find the problem the following procedure should get the job done. Obtain a variac–an adjustable auto transformer with a built-in current meter. Set it to zero volts and plug the amp into it. Very slowly, begin to increase the voltage and keep your eye on the current meter. As you increase the voltage, the current draw will skyrocket when the short kicks in. Back the voltage off until the current draw is just a couple of amps, which will probably be very close to zero volts. Now look for something that is getting hot. Whatever is overheating is where the problem is. This is a power-limited "smoke test." If done correctly, the problem can usually be found without damaging any expensive parts. It's a little crude, but effective. The variac can always be used during initial power up of any new project to protect against component burnout. I often use the procedure and it has proven to be very useful. Yes, I make wiring mistakes too.

If there are no shorts, then check to see if all of the output voltages are present. Be careful! Keep one hand in your pocket. The voltages are potentially lethal. They should all measure 10% to 20% higher than the rated levels because there is no load. The main thing we are testing for is to verify that voltage is there. So far, so good.

The next step is to install all of the tubes but **do not** install the output stage fuses. Strap an 8 ohm 20-watt resistor across the output. Radio Shack sells them. The circuit has to have a load connected for check out. Turn it on and carefully inspect all of the tubes. Keep looking for smoke. Allow a couple of minutes to pass. All of the filaments should be glowing. If any output tube is dark, then it has a short and is bad. If they are all glowing, proceed to the next test.

Measure the voltages for the circuitry as shown on the schematic. They should all be within 10% of stated ratings. I have never found an amp that had an audio problem that did not have a DC voltage problem. (These problems I refer to are not failures in the field, but wiring errors made during assembly.) Back to the output stage.

Set the bias adjust pot to mid rotation. Measure the bias voltage on each output tube. They should all be within a volt or two of each other. If any grid measures markedly lower than the other, then it is shorted. The positive tubes should measure about minus 40 volts. The voltage on the negative tubes is dependent on the position of the bias pot. Turn the pot from one extreme to the other and the voltage range should be from about minus 200 to minus 220 volts. Set it to about 210, shut the amp down, and install the output stage fuses. We are getting closer but are not out of the woods yet.

Turn the amp back on and after a couple minutes of warm up, set the speaker output to zero volts by adjusting the bias pot. If you can get to zero volts, actually anything within 50 millivolts is essentially "zero," we are probably there. Observe the reading. It should fluctuate slightly (variations of 10 to 15 millivolts) but if it changes wildly and won't calm down, then a tube is marginal and will fail shortly. The grid circuit is breaking down. When a grid shorts, the tube turns full on and the output voltage across the 8-ohm resistor will rise to 8 volts, positive or negative depending on which tube shorted. Don't forget to check the output stage fuses if an output tube starts sucking current. One may blow, which is what it is supposed to do. At any time during use, if you can't get the output to zero, check the output stage fuses first. Just because a fuse is blown, that doesn't mean a failure has occurred. It most probably means the amp experienced an intermittent overload and the fuse did its job. So back off the volume, please.

The output stage does not require matched tubes but they should be somewhat balanced. It is commonplace for the DC bias current through the tubes to vary by a factor of 3 to 1. If, by random chance, you get a couple of high current drawing tubes on one side and low on the other, the bias control will have to be skewed all the way to one end of its rotation and conceivably not even be able to reach a zero on the output. Ideally, the grid bias voltage on the negative tubes should be between minus 207 and minus 213 volts. If the bias voltage on the negative tubes is set to less than 205 or greater than 215 negative volts, the stage could stand for some balancing. The solution is to swap a tube from positive to negative bank to bring about a more balanced condition. The offending tube(s) can be found by measuring the voltage drops across the 2-ohm cathode resistors. The more equal the voltages are, the more balanced the output stage.

Having passed all of these tests, the amp should be ready to listen to. If you have access to an oscilloscope and signal generator, definitely run a sine wave signal through it and observe the trace on the scope. It should be clean and symmetrical. The final test is to drive the amp into clipping and try to pop any marginal tubes. That has only happened to me once or twice, but it has happened. When conducting maximum power tests, do not leave the amp in a maximum output condition for more than a few minutes at a time, or you'll risk overheating the output tube plates.

I don't mean to worry you with all of this failure stuff. I just want you to be aware of all the possible things that could go wrong–just in case something does go wrong. Then you will be able to recognize it and take corrective action instead of being stuck with a smoldering mess. The most common failure during initial checkout is the grid to cathode short. Odds are, you will be able to just plug in the tubes and everything will work perfectly. This gives you some insight into the care and effort that I, as a manufacturer, must give these amps to ensure the high reliability for which they have gained a reputation in the field. Nothing good ever comes easy.

The amp will sound a little harsh for the first hour of its life. It takes about 10 hours for most of the break in to take place. After about 50 hours, it is essentially through mellowing. Watch the bias setting for the first two weeks in case there are any infant mortality problems. Check the bias about once a month but you will find that adjustments are rarely needed.

PARTS LIST

R1	10
R2	100K
R3	698K
R4	220K
R5	3K
R6	100
R7	1M
R8	33K-2W
R9	39K-2W
R10	15K-3W
R11, R12	2K
R13, R14	300K
R15	2.7K-5W
R16	10K-2W
R17-R20	33K-2W
R21, R22	1K
R23-R26	100-5W
R27-R30	100K
R31-R34	2-5W
R35	30K
R36-R39	100-5W
R40	7.5-2W
R41-R44	100K
R45-R48	2-5W
R49	30K
R50	47K-2W
R51	1M
R52, R53	82K
R54	10K
R55, R56	47K-2W
R57	82-5W
P1	5K LINEAR POT, 1/2W, 500V
C1	22 MFD-450V
C2	100 MFD-25V
C3	.1 MFD-630V
C4	1500 PF-500V
C5, C6	47 MFD-350V
C7, C8	1 MFD-630V, QUALITY FILM TYPE
C9, C10	22 MFD-350V
C11	.1 MFD-630V
C12-C15	22 MFD-350V
C16, C17	.1 MFD-630V
C18-C21	2200 MFD-200V
C22	20,000 MFD-40V
C23	.01 MFD-400V
D1-D4	1A-1000V DIODE
Z1-Z3	150V-5W ZENER
B1	35A-600V BRIDGE RECTIFIER MUST HEATSINK
B2	35A-100V BRIDGE RECTIFIER MUST HEAT SINK
CL1	KEYSTONE CL40 THERMISTER
F1, F2	1.5 AMP SLOW BLOW FUSE
F3	6 AMP SLOW BLOW FUSE
S1	15A SWITCH
T1	200V-100MA TRANSFORMER HAMMOND 263AX
T2	120V-50MA TRANSFORMER
T3	240V-1.38A CENTER TAPPED TOROID, AVEL TRANSFORMER D4059
T4	20V-8A TOROID, HAMMOND 180 N40
V1	12AX7A VACUUM TUBE
V2, V3	12AU7A VACUUM TUBE
V4-V11	EL-509 VACUUM TUBE

FOR 25 WATT VERSION

ELIMINATE THE FOLLOWING:
V6, V7, V10, V11, R25, R26, R29, R30, R33, R34, R38, R39, R43, R44, R47, R48, C20, C21

CHANGE THE FOLLOWING:
F1, F2 TO 3/4A, SLOW BLOW
F3 TO 3A, SLOW BLOW
CL1 TO KEYSTONE CL60 THERMISTER
T3 TO 240V-.67A CENTER TAPPED TOROID HAMMOND 180 H240
T4 TO 20V-4.5A TOROID, HAMMOND 180 L40

PIN CONNECTIONS FOR EL-509

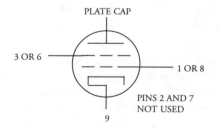

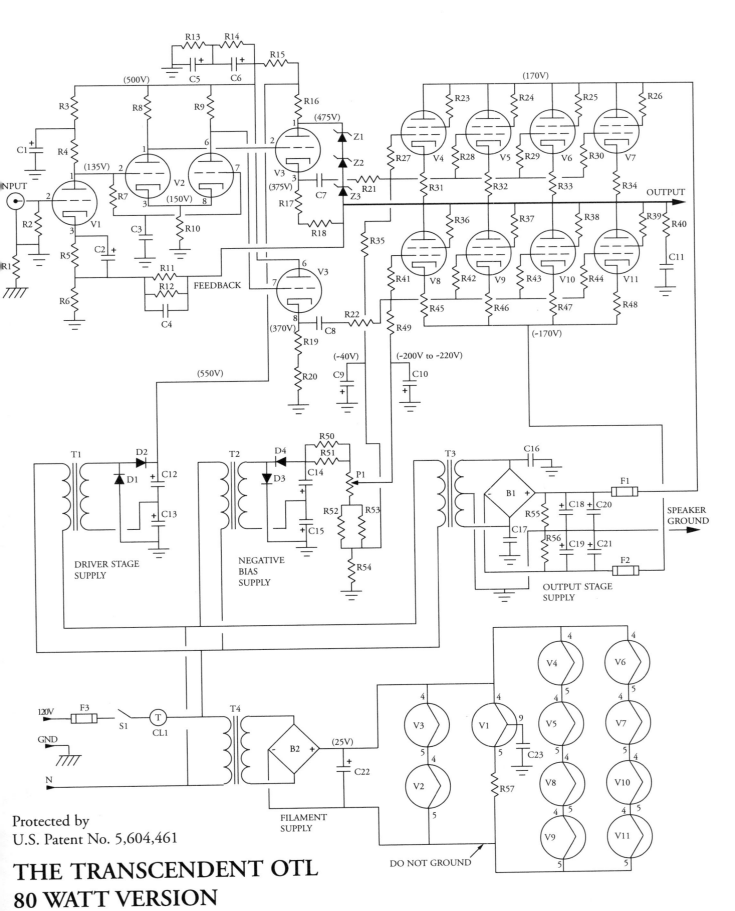

THE TRANSCENDENT OTL
80 WATT VERSION

Protected by
U.S. Patent No. 5,604,461

Copyright © 1999 by Bruce Rozenblit
All Rights Reserved

GROUNDED GRID PEAMP

My OTLs are very fast. If used with slower front-end equipment, their benefits will never be realized. There are many tube base line stages on the market that are terribly slow. In lieu of using these products, I often tell my customers that they would be better served with a high quality solid-state preamp instead.

There is absolutely no inherent reason why tube equipment has to have a slow, soft sound with the transients rounded off. To expand my product line and provide a good mate to the OTLs, I developed the grounded grid preamp.

Although this type of gain stage has been around forever, it has been virtually ignored by the industry. It was primarily used for radio frequency applications. Advantages are extreme bandwidth, very low noise, and high linearity. This sounded like a winner to me, so off I went to the shop with high hopes for a good design. I wasn't disappointed and you won't be either.

The bandwidth is flat from a few Hz to 300 kHz. Voltage gain is set to 12 dB which is just right for most applications. Distortion and noise are virtually nonexistent. Detail and resolution meet or exceed the best solid state preamps. Output impedance is low at 300 ohms, allowing it to drive anything. You can't overload it because it will output more than 20 volts RMS. It sounds good too–natural and musical with no coloration.

Power amps are expensive to make. They have large power supplies, transformers, and chassis which drive up the price. Preamplifiers do not have these requirements. Therefore, it seems reasonable to me that they should be much less expensive. Unfortunately, this is not often the case. This question should be asked before you buy anyone's preamp, including mine!

The design criterion for this project is a purist or minimalist design. I stripped out all of the features that everyone thinks they need and most people never use. This makes the project much easier to build and greatly lowers cost. There are no tone or balance controls, tape monitor loops, high and low-cut filters and, most important of all, no remote control. I have more remotes than dishes. The world is becoming enslaved by the dreaded remote control cartel. Lose one under the couch and your system is useless. Then, in a panic, you have to run all over town to find a replacement and are forced to pay the price. I often wonder if large electronics corporations have done studies of how much money they can make by predicting how many remotes will be lost.

I am sure that there are people who want one or all of these features but I targeted this product for simple playback from a few sources with the highest quality at the lowest cost. The mass marketers are probably laughing at me as they read this because such a product can't possibly sell. "Got to have gizmos and widgets," they would probably say. Well, we will see how it plays out over time.

When I finished the design and looked at the circuit, it struck me as remarkably similar to a solid-state differential amplifier. It was not my intent to mimic that topology with tubes, but that is what it took to make it work so well. I would not be surprised if many years ago solid-state engineers working on similar problems came to the same conclusions, hence the similarity.

A grounded grid gain stage works differently than a traditional gain section. All tubes are modulated by the voltage between the grid and cathode. In most applications, the cathode is held at ground and the grid is modulated. With the grounded grid, the grid is held at ground and the cathode is modulated, which is exactly the opposite. When this is done, the signal phase at the plate is *not* inverted as it always is for the grounded cathode. The problem here (there is always a problem) is that the input impedance for the stage is very low, restricting its applications. The solution is to buffer it with another stage that has a high impedance input and a low impedance output. The cathode follower fits these requirements perfectly.

Starting at the beginning, let's see how it all fits together. Both channels are identical so an examination of one tells the whole story. The input is a dual 50k audio taper pot. A standard pot is fine and you don't have to spend more than $5 for one. Expensive pots will track more closely through the minimum range of rotaion. The improvement is very slight and borders on the limits of perceptibility. I can't hear any difference. Some claim they can. If your ears are that sensitive and you listen to music at extremely low levels, then use a premium pot. I chose 50k because I thought it might be a trifle quieter than a 100k unit.

The decision to use a 50k pot may be a vestige of some residual audiophile neuroses that I still harbor. We all have some. I have managed to get rid of just about all of mine. Audio is a lot more fun when you finally become liberated from all that nonsense and just enjoy the music and the equipment. Some audiophiles are so terribly afflicted that they would be financially and psychologically better off to sell their equipment and just iron their socks. I have heard the most outrageous things like, "You can't touch tubes with your fingers," "You can't re-solder a connection," "You can't step on cables or you will realign their crystalline structure." Believe me, that's the mild stuff. It gets worse. Perhaps the remote control cartel has planted mind-altering chips in their products that brainwash people into believing anything they hear. Particularly if the message tells them to spend 10 times more on an item than it is worth, especially if it doesn't do anything. (Boy, am I going to get some hate mail!)

The signal then goes through R2, a 10k resistor that prevents small parasitic grid oscillations. The plate of the cathode follower, V1A, is held at about 75 volts. Its cathode is coupled to a 51k resistor and the cathode of the next triode, V1B. This is where the gain takes place. The audio input is developed across the 51k resistor because of the cathode follower action of V1A. Since the resistor is also connected to V1B, it is the input, or driving voltage, for gain stage V1B. Notice that the cathode resistor is rather large at 51k and is connected to a negative 200 volts. Remember, the negative 200-volt rail is at ground potential as far as the audio signal is concerned. For good linearity and efficient coupling between the stages, the cathode resistance needs to be large. To make room for the large voltage drop across the resistor, a negative power supply is necessary. Another alternative would be to use an FET as a constant current source. This would eliminate the need for the negative supply entirely. Other techniques are possible but they would require AC coupling the input, which is less desirable. The chosen method is totally reliable and has great sound, so I'm leaving things alone.

Instead of using a plate resistor as the load for V1B, I'm using another tube. This is an active plate load, a tube connected as a direct coupled constant current source. It doesn't have a high enough impedance to really function as a current source, but the connection is the same as an FET would have for that application. Its advantages are that it conserves power supply current, is fully direct coupled, and provides a low impedance output. With the values shown, it behaves like a plate load resistor of 63k which is just right for linear operation. The 10k resistor from its grid to the plate of the gain tube eliminates parasitic oscillations as on the input tube.

The output is taken directly from the cathode of the load tube and is AC coupled to the output jack. The grid of triode V1B opens up an interesting possibility. It must be referenced to ground to achieve a DC current balance through the tubes because the grid of the input tube is referenced to ground through the volume control. This is achieved by tying it to ground through a resistor. Because the output signal is in phase with the driving voltage, it can be applied to this resistor for negative feedback. The 100k resistor does just that and provides 7 dB of feedback. That is just enough to clean up the output and knock a little noise out of the signal. If you are being overcome with severe audiophile neuroses and are having a psychotic episode because I just "ruined" a perfect design by including a little feedback, then leave the 100k resistor out. I certainly wouldn't want to upset anyone. The gain will increase to 19dB and the output impedance will rise a little.

The filaments are powered from a 12-volt integrated circuit regulator for ultra low noise operation. It must be heatsinked. Notice that the supply is grounded. That is done to reference the 12-volt system to ground to eliminate any small parasitic oscillations between the filaments and cathodes. There is a tiny capacitance between those elements that can cause these to occur. We have to treat signals with a little more care in a preamp than a power amp because the lower amplitudes, render them more vulnerable to contamination.

The power supplies are very straight forward. Because of their low cost, it is very cost effective to use two small 120-volt transformers for the positive and negative supplies. Each has a rating of 50 mA. They are wired as voltage doublers. The zener diodes make an excellent low current shunt regulator just as in my OTLs. Extremely precise voltage regulation is not necessary. The shunt regulators hold the rail voltage steady if you pull the tubes out or the utility power starts fluctuating. The voltage in my shop often bounces between 115 and 125 volts. They also remove a great deal of ripple without having to resort to large, expensive capacitors. Many line stage preamps have huge power transformers just to supply a few milliamps of current. I guess they are selling their equipment by the pound. The fellow that wants the 90 pound single-ended amp must be on their customer list.

A line fuse is required to protect against component failure. All electrical devices should be fused for safety's sake. The .01 capacitor across the power switch is required to keep the unit from making a popping sound when you turn the unit off. There is no need for a thermistor to limit the current as the inrush with the small transformers is very low.

CHECKOUT

There is not a lot to test with this project. Insert all of the tubes and the line fuse and turn it on. If you hear any buzz at all from the transformers, there is a short. There should be no noise at all. Shut it down and see if a capacitor or diode is installed backwards. If all is quiet, measure the power supply voltages. The filaments should be glowing normally. Now measure the voltages on the schematic which demonstrates that the tubes are conducting and biased properly. If you have a signal generator, then input a sine wave and see that the volume control is not wired backwards. Clockwise rotation increases the resistance from the wiper to ground. Make sure the left and right channels did not get transposed somewhere along the way. There should be no DC voltage on the output.

There is one precaution when using this preamp with a solid state amplifier that responds down to nearly DC. At about five seconds into warm up, a voltage disturbance will appear on the output which will vanish in one second. This is caused by the output capacitor charging as the preamp warms up. The frequency of this disturbance is about 0.5 Hz, which is so low that it is practically DC. It has the potential to drive the amplifier's output stage full on for about one-half second resulting in a potentially damaging heavy current flow into your speaker. The risk of damage is dependent upon how rugged your speakers are and how your amp responds to these kinds of signals. The way to prevent any problems is to always turn the preamp on first, wait 15 seconds, and then turn on the power amp. Tube power amps pose no threat because they cannot pass signals of these extremely low frequencies. One way to engineer this away is to connect a muting relay with a 15-second timer that shorts the outputs to ground and then connects them to the circuit at the expiration of the timing cycle. Many commercial preamps employ circuits of this nature for the very same reason. It's time to hook it up to your system and have a listen!

PARTS LIST

R1	10
R2, R3	10K
R4	30K-2W
R5	10K
R6, R7	51K-2W
R8	3K
R9	100K
R10	20K
R11	10K
R12	3K
R13	100K
R14	20K
R15	5.1K-2W
R16	10K-2W
R17	5.1K-2W
R18	10K-2W
P1	DUAL POT, 50K, AUDIO TAPER
C1	22MF-350V
C2, C3	1.0MF-400V, quality film type
C4 - C11	22MF-350V
C12	.01MF-600V
C13	1000MF-25V
D1- D8	1A-1000V, 1N4007 DIODE
Z1-Z4	100V-5W ZENER DIODE
IC1	GENERIC 12V-1A REGULATOR
T1, T2	120V-50MA TRANSFORMER
T3	12V-1A TRANSFORMER
F1	1/2A, SLOW BLOW FUSE
S1	3A, SPST SWITCH
V1-V3	12AU7A VACUUM TUBE

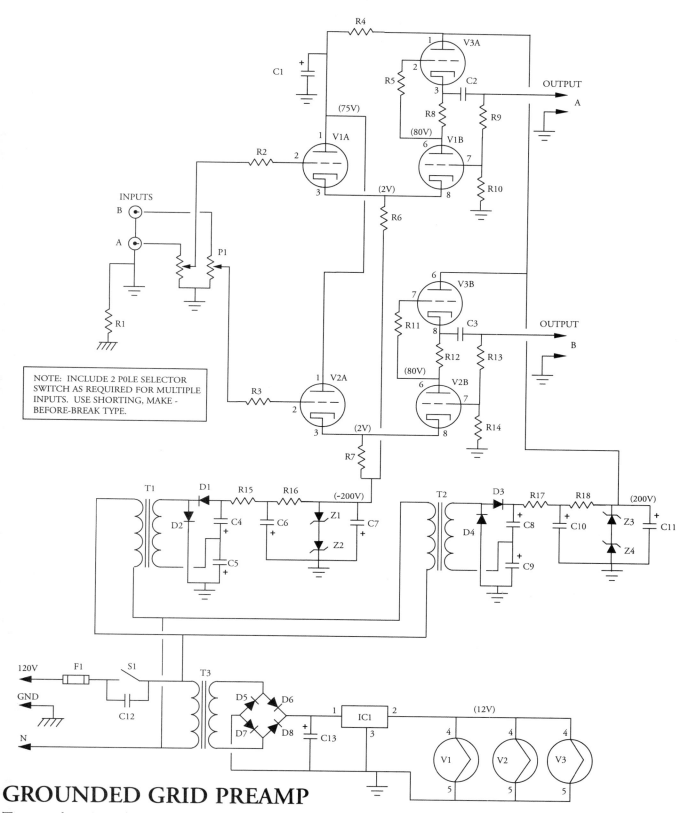

GROUNDED GRID PREAMP

Transcendent Sound, Inc.
Copyright © 1999 by Bruce Rozenblit
All Rights Reserved
Commercial Use Strictly Prohibited

SUPER COMPACT 150W AMP

Many people don't have a lot of surplus space that can be dedicated to an audio system. They need small components that can be placed on a bookshelf or installed in typical home-entertainment furniture. Some purists may disagree, but homes are for people first. The stereo system usually ranks two or three notches down on the priority list.

People do, however, want power. You can't have too many watts. High power OTLs are necessarily huge, which is why my largest model is 80 watts and not 250. Any larger and the thing just gets too big to move around, not to mention the problems associated with the additional heat load. So, I came up with the idea for an amp with a broader appeal.

Since the EL-509 can output more than one amp of peak current, why not take that same peak power ability and use it to make a super efficient transformer-coupled amp. No one else is, so it must be a good idea! This would give me the chance to see what I could accomplish in a more conventional format. The primary design goal was to create the smallest and lowest heat amp possible that could supply over 100 watts into a four ohm speaker. I wanted to develop an amp that could fit just about anywhere and drive just about any speaker. Most high-power tube amps have the proportions of a large microwave oven and can weigh over 100 pounds. You can get a hernia if you try to move one. That's hardly the type of product that can be easily placed in an out-of-the-way location.

Unlike the other projects in this book, the circuitry here is fairly conventional. It does, however, incorporate some new features and engineering refinements that improve performance and make it easy to live with. I didn't build a new house this time, I just gave it a fresh coat of paint.

The commercial model measures less than 10 inches square and weighs only 23 lbs. It will easily fit on a bookshelf. The little beast will make an actual 150 watts into 4 ohms. The heat rejection at idle is less than 100 watts for each channel. Talk about cool running! It will fit just about anywhere and drive just about anything. Because of the small size, it is an ideal candidate for bi-amping or home theater applications. By using off-the-shelf power transformers, your project will have to be a little larger, but it will still be very compact.

I never thought I could make a transformer coupled amp that sounded like this. It is not the least bit heavy and syrupy like most other tube amps. The sound is clear and fast, not overly warm and slow. It sounds remarkably similar to the OTL, which should be no surprise to anyone since the same guy designed it (me). The high frequency half-power point (-3 dB) is at 50 kHz. High frequency transients are crisp and realistic. Other than that, the compact amp's most prominent sonic characteristic is a pounding, thundering bottom end. As I said before, there is no alternative to current drive when it comes to the bass.

The amp is 3 dB down at 25 Hz and then falls off rather rapidly. There is virtually no musical energy below 30 Hz and very little below 40 Hz. I did not see the point of trying to extend the bottom end into the subsonic region. That would require a monster of an output transformer and greatly increase the cost. The transformer used is a standard off-the-shelf unit that is sold by many parts distributors. I didn't want the design to be based on any crucial, custom-made parts. If you want to hear dinosaurs stomping through your living room, then an active sub-woofer will be necessary anyway. My concept of a home theater application would be to use the compact amp for the left, right, and center channels, a mid sized (50-watt) solid state amp for the rear, and an active sub-woofer for the earthquakes and such. That should sound pretty good! As far as a more traditional music application is concerned, this amp will definitely rock and roll!

The output impedance is 0.6 ohm, which provides for very good damping. The forward voltage gain is 24 dB, making the amp very easy to drive from virtually any source. Noise performance is excellent at greater than 90 dBA. The commercial unit is dead quiet. The ground plane cicuit board provides best possible noise performance. In fact, this is the quietest amp in the book. This noise thing can get out of hand. I've had customers call me and tell me they bought a noisy OTL. I ask them how far away their ear was from the speaker when they noticed the noise and they say, "Up against the grill." I tell them to forget about it. If you want a lot of high quality, quiet, tube power in a small space, this is the obvious choice.

I take great pride in bringing to market radically new circuitry, but in this case no new circuitry was needed. I could have complicated the design in order to claim some "breakthrough" discovery but that would make me a hypocrite, the cost would go up, and reliability would be diminished. I'm sure you've heard the old saying, "If it isn't broken, don't fix it."

The first two stages are very similar to the ones in the OTL. The biggest difference is that the second half of the input tube is used as a cathode follower buffer between it and the phase splitter. It was there, so I used it. The only reason I didn't use the buffer in the OTL is that in the commercial stereo amp, the second half of the input tube is used for the other channel.

The only other major difference is that the cathode resistor for the input stage is not bypassed with a capacitor. This was done to reduce the gain and increase stability. When the resistor was bypassed, gain increased to 32 dB (feedback was set at 20 dB). That's just a little too hot for most signal sources. The increased stability with the reduced gain is definitely a plus. The amp is completely stable with a 2.2 mfd load driven by a 10 kHz square wave. Besides, the way the amp is shown, it sounds really, really good. When I get to the "really, really good" stage, I stop changing things.

The output tubes are connected in full tetrode mode. This greatly reduces the cost of the power supply as extensive filtering is not required. The current through a pentode or tetrode is mostly determined by the screen and control grid voltage. It is insensitive to plate voltage over a wide range. The range of insensitivity is much greater than the power supply ripple, rendering it benign. A common argument for the use of massive filter banks is that they increase peak power capabilities. But if you have 150 watts to work with, what do you need peak power for?

The output transformer is configured for a 4-ohm secondary. Most people have 4-ohm speakers and need the current drive. The impedance reflected to the plates will double when using an 8-ohm speaker which reduces distortion. The power supply voltage is very high at 700 volts giving plenty of room for the larger primary amplitudes that the higher impedance 8-ohm speaker requires.

Speaking of voltages, the 700 volt output stage supply is extremely dangerous. Not only is the voltage bad news, but it can generate one amp of current which is enough to kill several people. So, exercise the utmost in care when conducting tests. You certainly wouldn't use a chain saw to give someone a hair cut. Likewise you wouldn't troubleshoot this amp with wet hands. If you think I'm trying to scare you, you're right! If you're scared, you're safe. Accidents happen when people become complacent and careless.

The screen grids are held at 150 volts. This voltage must be clean and stable for low noise operation and good control of bias current. The best way to accomplish this is with a voltage regulator. The screens draw only a few milliamps but they are powered from a 700 volt source. The most reliable way to implement a regulator in this environment is with another tube. One of the main reasons many manufacturers make unreliable tube equipment is that they hang solid-state regulators all over their circuits which were never intended to function at such high voltage levels. The regulator for this project is a very interesting tube, the 6BM8. It is a dual tube containing both a high gain triode and a low power pentode. I am using it in an unusual configuration. Most regulators use a low impedance triode as the series pass element but I chose a pentode which has a higher plate impedance. The regulator's output impedance is still low because the pentode configuration provides gain which eventually goes into the negative feedback of the regulator and ultimately reduces the output impedance. The pentode must drop 550 volts across it. Fortunately, the 6BM8 pentode has a 600-volt rating, a major reason why I chose it. Plate dissipation is 7 watts which is adequate for this application. The amplifier element for the regulator is the high gain triode section. It is referenced to a 51 volt zener diode and compares the output voltage to the reference via sampling resistors R35 and R37 which constitute a negative feedback circuit. Any voltage difference between the sampled voltage and reference is amplified by the triode and applied to the pentode in the direction that reduces the sampled differential.

For a pentode to act like a pentode, its screen grid must be bypassed to its cathode. That's the function of C17. It can be small because the screen resistors are large. The 33k resistor on the output is needed to give the regulator a minimum load to work into in order to bias the pentode. Without it the triode will push the pentode essentially into cutoff, a turned-off state with no current flowing through it. The tube makes a very simple, reliable regulator that gets the job done.

The cathodes of the output tubes are each grounded through a 10-ohm, 5-watt resistor. They are integral to the metering scheme and provide a degree of degenerative feedback during high current operation where it is needed most. Here goes another myth down the drain. There is no precise bias current setting for optimum operation. There are only ranges of operation. Generally speaking, the heavier the bias current, the warmer the tube will sound. Some people think the warmer the better. I definitely do not. My designs are based upon maximizing accuracy. What is critical here is not the absolute level of current, but the *balance* of current. Output transformers cannot tolerate much of a DC current imbalance. Too much imbalance will cause premature saturation and severe distortion.

If the currents through the 10-ohm cathode resistors are equal, the voltages at the cathodes will be equal. The voltages are compared by inserting the test probes of a digital voltmeter into two pin jacks which should be mounted on the front panel. By obtaining a zero volt reading on the meter, an accurate balance of current is achieved. The commercial unit employs a panel meter.

The balance is obtained in a novel way. The bias is set by using a dual pot that is crosswired, meaning that the input terminal on one section is the output terminal on the other. As the pot is rotated in one direction, the voltage rises on one wiper and falls on the other. The opposite happens if the rotation is reversed. So, when making a bias adjustment, the bias voltage is increased on the tube that is drawing too much current, thereby reducing it. Simultaneously, the bias voltage is reduced on the tube that is drawing too little current, thereby increasing it. The difference in current draw between the two tubes is reduced to zero as they are both pulled toward each other. They meet somewhere in the middle. I am extremely pleased with the simplicity and versatility of this adjustment arrangement and plan to standardize it for use in any future transformer output amps that require a bias adjustment.

The output tubes make great radio

frequency amplifiers so care must be taken to keep them from oscillating. A 1k resistor must be connected directly to the socket pin of the screen grid, and a 20k grid resistor must be connected to the socket pin of the control grid. Likewise, the 100k resistors connecting to the bias pot must also be connected to the socket pin of the control grid. By doing so, the circuit is made immune from stray capacitances and is very stable. Circuits for audio frequencies that have trouble maintaining stability with respect to wiring and component placement are probably in need of some redesign. Physical placement should not be that critical at these frequencies. Circuits that operate well into the megahertz range have much different requirements and component placement can be critical.

There are two stability networks. The first is a "shelving" network which is R7 and C2 connected across the plate resistor of V1A. It starts attenuating the open loop response at an ultrasonic frequency and then levels off, hence the term "shelf." This provides a needed reduction in open loop gain to stabilize the amp when the feedback is applied. The second network is the 1500-picofarad capacitor across the 2000-ohm feedback resistance. This network strongly affects the rise time and damping of the amplifier, and sets the amount of feedback. I prefer to slightly over-damp to prevent any high frequency or ultrasonic oscillations from occurring. I am using 17 dB of feedback. Transformer coupled amps are much harder to stabilize because of the reactance of the output transformer. Don't even think about changing the values of any of the components in the compensation networks.

The power supply uses three transformers. We could get away with one, but performance would suffer. There are stacked core-type transformers available but I've found them to be unsatisfactory in this application. Three major problems appeared. One occurs when the power supply rectifiers switch. Harmonics are generated (spikes) that cause the casing to vibrate and even make a whistling noise. This is hardly acceptable unless the amp stays in the basement. The second problem is that the spikes capacitively couple to the filament winding and are so strong that they blast through to the cathodes and end up in the speaker causing a low-level, midrange buzz. The third problem is that the transformer generates strong magnetic fields that couple to the output transformer and put a lot of hum into the speaker. Rotating the transformer core 90 degrees could not eliminate it. So the wrong kind of transformer can cause three different kinds of noise.

Back to the toroid we go. I find it ironic that the classical application for toroids is in preamplification because of their low magnetic fields; however, I use them in power amps and use stacked core transformers in my preamps. There I go doing the "wrong" things again. The reason this works is that the preamps have very small transformers which generate small magnetic fields. There's lots of room in the preamp to mount the transformers far away form any sensitive areas, and there aren't any magnetic components in the preamp, such as an input transformer, which are susceptible to external magnetic fields. Power amps, on the other hand, require large transformers that generate strong magnetic fields. There is also a highly susceptible magnetic component sitting right next to it, the output transformer. The transformers are so big, it is next to impossible to mount them far enough away from each other to eliminate the interaction. Then we have the spikes coupling to the filament winding. So many problems must be overcome.

The most economical way for the hobbyist to get the right transformers is to use standard, mass produced units. The output stage needs a 230 or 240 volt secondary, rated for at least 0.75 amp. The filaments will require a 7 amps at 6.3 volts, while the bias needs a small 120 volt center tapped, or 60 volt, 50 milliamp unit. The output stage transformer will be used as a voltage doubler. This is a much more cost-effective use of the transformer. A small capacitor (C8) is connected across the secondary to absorb voltage transients to protect the rectifiers. In order to make the power supply super tough, I used two rectifiers in series to double the voltage rating as high amperage 2000 volt rectifiers are not readily available. The one megaohm resistors across them serve to equalize the voltages. A common failure in contemporary tube amps is loss of a power supply rectifier. By adding two more (50 cents) and a capacitor (60 cents) and 20 cents worth of resistors, the power supply should be highly resistant to failure. You would be surprised how many high dollar amps don't do this. I guess if your enclosure is beautiful, customers don't mind frequent equipment failures.

The filament supply is AC. It has 100 volts DC impressed upon it to relieve the stress between the cathodes and filaments. This voltage also has the effect of "sealing" off AC leakage between the filaments and cathodes, thereby reducing noise.

CHECKOUT

Install all of the tubes and fuses. Position the unit so you can observe the tubes. Turn it on and watch. If a tube arcs over, you will see a purple flash. I'd get another tube. All filaments should be glowing. Standard power amp check out rules apply. If you hear a loud transformer buzz, see or smell smoke, shut it down. If everything looks all right, turn it off and position it on its side so you have access to the circuitry. Connect the ground of your voltmeter to circuit ground with an alligator clip so your hands are free of the chassis. Turn it on and step back a little. You do have a strong light illuminating the chassis so you can see what you are doing, don't you? Take a minute to watch and see if any components are overheating or there are any other problems.

Now here is the dangerous part. Take the voltmeter test lead in one hand. Put your other hand in your pocket! DO NOT allow any part of your body to come into contact with the chassis and circuitry. Measure the power supply voltage at the output transformer. It should be around 700 volts. Now measure the voltages for the phase splitter and input stages. Check the voltages at the grids of the output tubes. They should be in the mid-negative forties. The screen grids should be at 150 volts.

After several minutes of warm up, balance the output tubes. They need several minutes to settle down. The bias setting should be checked again in an hour. You will find that there is very little fluctuation. Don't worry about trying to get a super-accurate setting. It's just not that critical. There really isn't anything else to check.

PARTS LIST

R1	10
R2	100K
R3	200K
R4	3K
R5	100
R6	2K
R7	10K
R8	100K
R9	200K
R10	33K-2W
R11	39K-2W
R12	15K-3W
R13	1M
R14, R15	20K
R16, R17	100K
R18, R19	10-5W
R20, R21	1K
R22	15K
R23-R26	1M
R27	15K-3W
R28-R31	300K
R32	200K
R33	100K
R34	10K
R35, R36	200K
R37, R38	100K
R39	33K-2W
R40, R41	100
R42, R43	51K

P1	DUAL POT, 10K-1/2W, LINEAR, 500V

C1	22MFD-450V
C2	50PF-500V
C3	1500PF-500V
C4, C5	.1MFD-600V, QUALITY FILM TYPE
C6	.1MFD-600V
C7	22MFD-160V
C8	.01MFD-800V
C9-C14	100MFD-400V
C15, C16	22MFD-350V
C17	.1MFD-630V
C18	22MFD-160V

D1, D2	1A-1000V DIODE
D3-D6	3A-1000V DIODE
Z1, Z2	51V-1W ZENER

F1	5A, SLOW BLOW FUSE
CL1	KEYSTONE CL60 THERMISTER
S1	10 AMP SWITCH

T1	120V-50MA, CENTER TAPPED TRANSFORMER: OR 60V UNIT WITH FULL WAVE BRIDGE
T2	240V-.94A TOROID, HAMMOND 180 J240
T3	6.3V-8A, TRANSFORMER.
T4	OUTPUT TRANSFORMER HAMMOND 1650T

V1	12AX7A VACUUM TUBE
V2	12AU7A VACUUM TUBE
V3, V4	EL-509 VACUUM TUBE
V5	6BM8 VACUUM TUBE

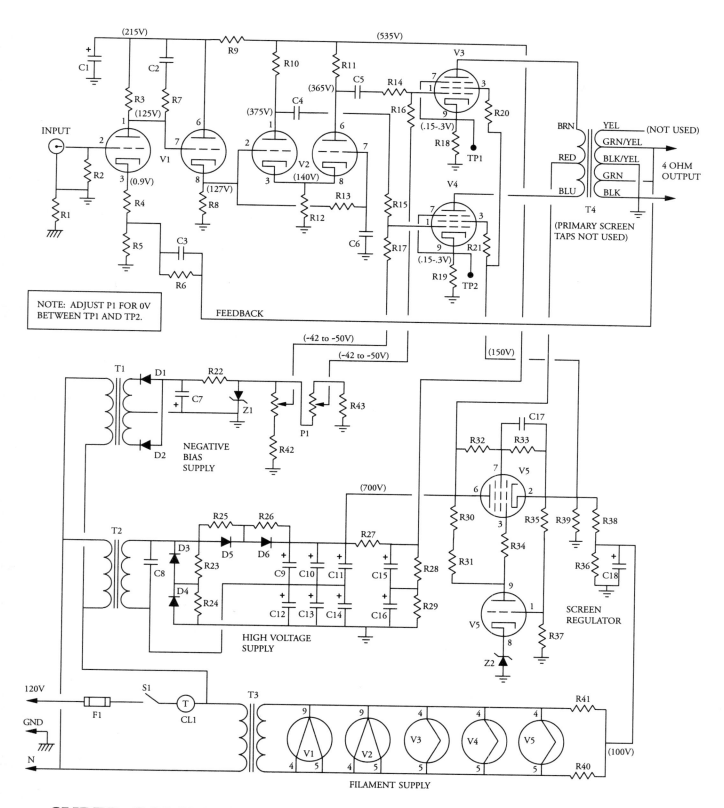

SUPER COMPACT 150W AMP

Transcendent Sound, Inc.
Copyright © 1999 by Bruce Rozenblit
All Rights Reserved
Commercial Use Strictly Prohibited

SINGLE ENDED WITH SLAM

I never really understood the big commotion about single-ended amps. Why would anyone want an expensive amp that has very low power, no bass, a restricted top end, and high distortion? Apparently I must be missing something because these things have been big sellers. After many conversations with audiophiles and studying charts of distortion spectra for several single-ended amps, I think I have finally figured out what the attraction is.

The characteristic of single-ended amps that audiophiles tell me they prefer most is the way vocals are reproduced. An almost enchanting, mesmerizing effect is usually described. This effect is so desirable and powerful that people are willing to overlook the shortcomings of the amp. What causes this reaction?

By studying charts of distortion spectra for several single-ended amps, I have concluded that the desired effect is probably produced by large amounts of even-order harmonic distortion. The distribution of the distortion is an array of even-order harmonics that linearly decline with each octave. This produces a euphonic effect that is psychologically or emotionally pleasing to some listeners. I would liken the effect to vocalists singing in harmony. The interaction of the singers' harmonics produces a pleasing sound we call *harmonizing*. Vocals obtain a richness and fullness that otherwise would be absent. Therefore the classic single-ended amp provides a particular coloration that some individuals find extremely pleasing. I have no objection to this decision and have already stated that the only person the audio system has to please is the person using it.

With my curiosity piqued, I had to find out for myself what single-ended sound was all about. I set out to build an amp to experience the phenomenon first hand. (You certainly didn't expect me to buy one, did you?) I designed and built a traditional single-ended amp with no feedback using an output tube with a directly heated cathode. I wanted to listen to it before any extensive measurements were made.

The sound was very smooth and pleasing. There was a loss of detail but the character of the sound was lush and desirable. Treble response was quite good and I didn't notice any serious loss of high-frequency information. Bass response was pathetic. It just wasn't there. Damping factor was so poor that transient bass tones, such as from an electric guitar, lost all percussiveness and sounded like continuous tones that just got louder and softer. I actually couldn't stand to listen to it.

Measurements revealed why the amp sounded as it did. Output impedance was terribly high at 3.5 ohms. No amplifier with an output impedance that high can accurately drive a woofer. Power output was 12 watts, which is a reasonable level. Conducting frequency sweeps demonstrated how the power output dramatically dropped below 100 Hz. At those frequencies, the signal waveform became highly misshapen and distorted. The distortion increased as the frequency dropped. The amplifier simply could not handle low-frequency signals.

The top end was actually quite good. The signal waveform was clean and extended well beyond 20 kHz. It seems as though this amp would be ideal for use in a biamplified subwoofer-satellite system.

Harmonic distortion wasn't that high at 3% at 10 watts. It dropped considerably as power was reduced. At one watt, it was only 0.5%. I expected the distortion to be much higher. Perhaps units that generate more distortion impart even more lushness to the sound.

The character of the sound had intrigued me to the extent that I wanted to investigate the concept further. The nature of the sound (above 100 Hz) was definitely very enjoyable to listen to. I pondered that if the coloration could be toned down a bit and the amp had a solid bottom end, this could become a significant new design. I knew from the start of the project that the biggest obstacle to overcome would be the bass. Analyzing the electrical behavior of the output stage should reveal the right direction to pursue in solving the problem.

The primary winding of a single-ended transformer is highly inductive. Therefore, it must consume a lot of reactive power. The impedance of an inductance decreases with frequency. Consequently the lowest frequencies will require the greatest amount of reactive current which further worsens bass linearity. The best way to drive a highly reactive load is with a low impedance, high current source. A low impedance source should be better able to control the nonlinear reactive load. The principle is the same as an amplifier with a high damping factor being able to better control a woofer.

The lowest impedance configuration for a vacuum tube is the cathode-follower. If my assumptions were correct, then driving the single-ended transformer with a cathode-follower output stage should make for a significant improvement. It most certainly did.

Cathode-follower output stages are extremely difficult to design and this project was no exception. The final circuit turned out to be somewhat complicated. One factor that probably slowed me down was my minimal research into existing designs. I wanted to embark on a journey of original discovery and not have any preconceived notions about how the amp should be designed. The school of hard knocks is a terrific place to get an education. I'm glad that I took this approach because I learned many new things. It's so refreshing and rewarding to develop new designs in a field where the vast majority of participants believe that "it has all been done before" and there are no frontiers left to explore.

Cathode-follower output stages require enormous drive voltages. Power supply voltages can easily reach the kilovolt level. Utilizing such huge voltages will greatly increase the cost of the project and make it exceptionally difficult to build. Hardware requirements and assembly techniques are much more stringent at these voltages. One way around this dilemma is to use a special custom-wound output transformer with a very low primary impedance on the order of only 200 ohms. Then you just pile on the output tubes in order to generate sufficient current to drive the primary. I did not want to design a project that used custom parts which are impossible for the hobbyist to find. It is also more cost effective for my company to use standard parts instead of custom-made ones because of economies of scale.

The selected transformer has a primary impedance of 1600 ohms. It also has the lowest retail price of any single-ended transformer on the market, which has as much to do with my choice as anything else. The impedance rating of the primary corresponds to using an 8-ohm load on the 8-ohm secondary tap. By connecting an 8-ohm load to the 16-ohm secondary tap, the reflected primary impedance is reduced to 800 ohms. Similarly, a 4-ohm load would be connected to the 8-ohm tap. The altered connection scheme permits using drive voltages that otherwise would have to be much higher.

The next step in the design process was choosing an output tube and corresponding power supply voltage. I didn't see the point in using exotic output tubes that cost hundreds of dollars for obvious reasons. Triodes are preferred because of their low noise and distortion. Also, the marketplace demands triodes in single-ended designs. One of the most rugged, low-cost, widely available, high-current output tubes is the venerable 6550. Although this is a beam pentode it can, like all other pentodes, be connected for triode operation. The maximum plate supply rating is 600 volts which is practical to work with. Tubes with higher plate voltage ratings tend to be very expensive and have higher plate impedances. So the choice was made to use the 6550 with a 550-volt supply.

The plate impedance of the 6550 in triode mode measures about 700 ohms. Triode operation ties the screen grid to the plate. The screen has a voltage rating of only 400 volts, which obviously presents a problem with a 550-volt supply. My solution was to connect the screen to the plate through a 150 volt, 5 watt zener diode. The voltage drop across the diode will keep the screen voltage at a safe level while still maintaining triode operation. The plate impedance for this configuration was found to be only 500 ohms. This was an unexpected benefit. I have not yet been able to determine why the plate impedance dropped. The effect must be tied to the action of the tubes electric fields.

The output impedance of a cathode follower is roughly calculated by dividing the plate impedance by the amplification factor. In this case, the output impedance calculates to be only about 65 ohms. The design uses three 6650s in parallel for a total output impedance of only 23 ohms. Each tube can supply about 250 milliamps. The output stage then has a current capacity of 750 milliamps, an output impedance of 23 ohms, and a total combined plate dissipation of 126 watts. That should be enough horsepower to drive the transformer primary.

The current demands placed on the output stage are much higher than the power requirements indicate. This is because the primary is highly reactive and consumes additional reactive current. Also, the reactive impedance of the loudspeaker is reflected back into the primary, further loading it down. That's why it is crucial for the driver of a single-ended transformer to be able to supply several times more current than what simple power calculations indicate. Tests of the cathode follower output with no other ancillary circuitry showed extremely linear operation at frequencies below 100 Hz. The significance of this is great as I used the *same transformer* in the cathode follower as I did in the experimental traditional single-ended amp. The dramatic improvement in low frequency performance with the same transformer verifies my claims regarding the reactive current requirements of a single-ended transformer output stage.

With the primary winding connected to the cathodes, the output stage experiences 100% degenerative feedback. This provides highly linear operation and is responsible for greatly lowering distortion. However, the degeneration makes the stage much harder to drive as it requires huge drive voltages to overcome it. I had to come up with a way to overcome this obstacle. Otherwise, power output would be limited to only two or three watts.

What I ended up doing was bootstrapping another cathode follower to drive the output stage. The signal requirements of the bootstrapped driver are even greater than the output stage. It needs to have even more voltage across it to create an envelope large enough to contain the necessary drive signal. Because there are already two capacitive couplings in the circuit, this stage needed to be direct coupled to the output to ensure good stability.

By using a negative 150-volt shunt power supply to connect the cathode resistor of the driver stage to the transformer primary, the DC voltage applied to the grids of the output tubes can be made as negative as needed to properly bias the output stage, thereby eliminating the need for a coupling capacitor. The negative supply also provides additional voltage headroom in which the driver can operate. This eliminates the need for an even higher plate voltage supply.

The bias voltage for the output stage is the combination of the drop across the resistance of the primary winding (about 27 volts) subtracted from the negative voltage at the cathode of the driver (about minus 18 volts). The total bias voltage is then minus 45 volts. The bias current setting for the output stage is set by R19 at 2200 ohms which adjusts the current through the driver tube. The greater the current, the more positive the bias voltage becomes. The current through the transformer should be set to about 150 milliamps (50 milliamps per tube) which is determined by measuring the voltage across the transformer primary.

The connection of the shunt regulator (zener diode Z1) to the primary, provides for the bootstrapping of the driver stage. Any signal developed across the primary is added to the signal developed at the cathode of the driver stage. The two signals are essentially equal which overcomes the 100% degenerative feedback of the output stage. Think of it as: one plus one minus one equals one. The circuit's function is almost identical to the floating cathode follower used in my OTL.

The tube used for the driver is the pentode section of the 6BM8. It has a 600-volt rating and the ability to withstand signal peaks that reach 900 volts. It is connected as a pentode cathode follower. The triode section is unused as it is unsuitable for this project.

Even with the bootstrapped arrangement, the signal requirements of the output stage are extreme. An audio signal voltage of 400 volts peak-to-peak is needed to drive the output to 10 watts. This is twice what the OTL requires. As you might have guessed, developing that much signal voltage isn't the easiest thing in the world to do. The signal must be clean with low distortion. A total of three triodes were used to boost the signal to the necessary levels.

With no feedback, output impedance measured 2.8 ohms. That is still too high to accurately drive a woofer. Negative feedback must then be applied to reduce the output impedance. I chose a damping factor of 10 which requires the output impedance to be no more than 0.8 ohm. Dividing 2.8 by .8 gives a reduction factor of 3.5 times. The amount of feedback that will achieve this much reduction is 11 dB. The final design uses 10 dB of feedback and the output impedance measures 0.9 ohm. That's close enough.

One of the problems encountered in the design of this project was getting enough gain and voltage swing at the same time. Tubes that provide high gain can't generate enough signal voltage. Tubes that can generate large signal voltages have low gain. The choice of tubes used was primarily made to solve this problem. The first stage is one section of a high mu 6SL7. Its output is coupled through a cathode follower formed from the other 6SL7 section. The follower is necessary because, at these extremely large signal voltages, the bias of the high gain tube is disturbed causing premature unsymmetrical clipping. The final gain stage is one section of a medium mu 6SN7. It can provide the huge 400 volt peak-to-peak voltage swing.

Because of the many gain stages, several damping networks were needed to stabilize the circuit. The first stage suffered from parasitic oscillations. A low pass network comprising R3 and C1 was necessary to eliminate the problem. Also, a series network of R4 and C2 was needed across plate resistor R5. Capacitor C9 installed across feedback resistor R33 minimized any overshoot and ringing of the amplifier when square wave testing was done. I always compensate an amp to minimize ringing at the expense of bandwidth because I feel it sounds better this way. High frequency harshness is very irritating. Each stage has a 10K grid resistor. Lastly, 10K grid resistors were installed in the grid circuits of each output tube to keep them stable and prevent any excessive grid current. Leaving out or altering any one of these stability networks will cause severe problems. I had a difficult time calming this circuit down. Each stage required some stabilization work.

The power supply is straightforward and does not incorporate any regulation. Although the commercial unit uses a toroid transformer, good results can be obtained with a standard stacked core type. The output transformer, either because of internal shielding or its winding configuration, could be placed near the power transformer and not pick up unacceptable amounts of hum. This was not the case with the 150W compact amp. For minimum inductive coupling, mount the power and output transformers at opposite ends of the chassis.

Rectifier noise capacitively coupled to the filament winding is still a problem. For the lowest noise performance, a separate six amp filament transformer should be used for the output tubes and bootstrapped driver. The center tap of the filament supply for the outputs and driver is connected to the transformer primary to relieve any voltage stress between the filaments and cathodes and also to reduce hum. The cathodes of these tubes operate at the 400 volt peak-to-peak drive voltage which causes the voltage stress. The voltage gain tubes should not be subjected to this voltage stress. Another one-amp filament transformer is used for the two gain stage tubes because it has to be biased up to 100 volts to minimize heater-cathode voltage stress. All of the filaments are AC. There is virtually no 60 Hz noise in the output so I felt that there was no need to go to the expense of using DC filaments. In the commercial unit, all three secondary windings are resident in one transformer. Once again the toroid proves to be the best choice for tube applications.

The triode output is much more susceptible to power supply ripple mandating good filtering. I used a choke/capacitor type filter with two stages of filtering. This subdued the ripple to very low levels. The amplifier is remarkably quiet without having to result to massive, expensive filtering sections.

Output stage bias current should remain stable because of the degenerative feedback caused by the primary's resistance. The more current that flows through the output stage, the higher the bias voltage. This has the effect of reducing the flow of current. The direct coupled driver has no large resistance shunted to ground which would create a positive grid voltage if the output tubes generated grid leakage current. Therefore, the inherent configuration of the output stage is maintenance free. Matched output tubes are not required either. It's a great benefit to the consumer to design tube audio equipment that is just plugged in and used, day after day with no adjustments required.

So what's the payoff? How does it sound? Bass response is only 1 dB down at 20Hz! That is nothing short of amazing performance for a single-ended amp. The bottom end is solid and not the least bit abbreviated. Bass transients are clean and sharp without any overhang. Of course only 10 watts are available so the bass pulses aren't going to give your internal organs a massage, but the foundation is there. The midrange and treble capture that almost magical single-ended quality. Although, it's not as heavy or overbearing. The sound is still extremely smooth and lush but retains more crispness than a conventional single-ended amp. You can definitely tell that you are listening to a single-ended amplifier. The distortion measures 0.5% at one watt, 1% at 3 watts and hits a maximum of 1.4% just before clipping. Again, this is very clean performance for this type of amplifier. The top end is extended and the amp is very quick. The treble is only 3 dB down at 35 kHz. Voltage gain is 21 dB allowing it to be driven from most any source. Square wave response at 100 Hz exhibits very little tilt, indicating excellent low frequency bandwidth and phase linearity. At 1 kHz, square waves look about perfect. At 10 kHz there is a slight rounding of the corners and some minor vertical slope, showing that the test frequency is close to upper bandwidth limit. However, there is absolutely no overshoot and ringing which verifies its superb damping characteristics. The amp is completely stable driving a 2.2 mfd. capacitor at 10 kHz.

I like it! I would have to say that some coloration is added. But maybe that's a good thing, like adding seasoning to a recipe. Proper seasoning can bring out the best flavors in food. Have I become single-ended crazy? Since I've fixed the problems inherent to traditional single-ended amps, my conversion is a distinct possibility. This is a terribly tough call to make. I now have three amplifiers and thoroughly enjoy all of them even though they each have a distinctly different sound. I'm going to have to add another rack and use all three. Oh well, worse things could happen.

CHECKOUT

The checkout concerns only making sure the power supply voltages are correct and a few DC levels are in the correct ranges to indicate proper tube bias. The most critical voltage to check is the DC voltage at the output transformer primary. It should be around 25-29 volts. The voltage at the output tube grids should be around minus 18 volts.

Check the voltages at the points shown on the schematic. The tubes will need a few hours to stabilize. Now all you have to do is show the amp off to your friends and let them guess where the bass is coming from.

PARTS LIST

R1	10
R2	100K
R3, R4	10K
R5, R6	100K
R7	1K
R8	100
R9, R10	300K
R11	39K-2W
R12, R13	30K-2W
R14	220K
R15	10K
R16	3K
R17	10K
R18	100K
R19	2.2K
R20	13K-2W
R21, R22	30K-2W
R23	200K
R24	100K
R25	3K-2W
R26-R29	300K
R30-R32	10K
R33	1.5K
R34-R39	300K
R40, R41	100
R42	100K
R43, R44	200K
C1	250PF-500V
C2	100PF-500V
C3	100MFD-25V
C4, C5	22MFD-350V
C6, C7	.1MFD-600V, QUALITY FILM TYPE
C8	100MFD-25V
C9	1500PF
C10	.1MFD-630V
C11-C14	22MFD-350V
C15	.01MFD-800V
C16-C23	100MFD-350V
C24	22MFD-160V
D1-D4	1A-1000V DIODE
Z1-Z4	150V-5W ZENER
F1	2A SLOW BLOW FUSE
S1	6A SWITCH
T1	OUTPUT TRANSFORMER ONE ELECTRON UBT-1
T2	400-0-400, 200MA TRANSFORMER HAMMOND 278X
T3	6.3V-1A TRANSFORMER
T4	6.3V-6A CENTER TAPPED TRANS.
L1, L2	1.5H-200MA CHOKE
V1	6SL7 VACUUM TUBE
V2	6SN7 VACUUM TUBE
V3	6BM8 VACUUM TUBE
V4-V6	6550 VACUUM TUBE

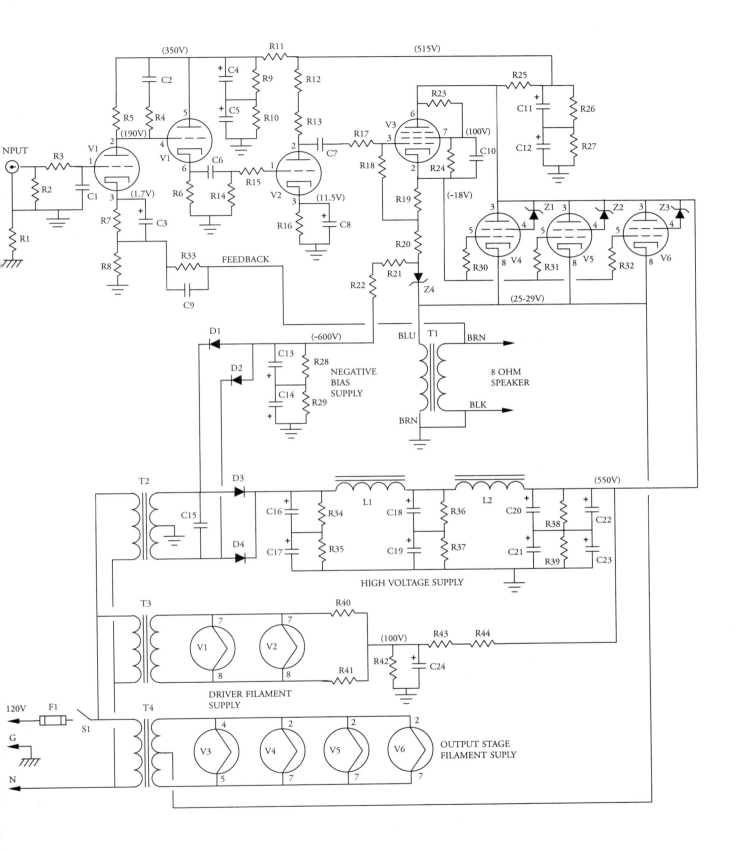

SINGLE-ENDED WITH SLAM

Transcendent Sound, Inc.
Copyright © 1999 by Bruce Rozenblit
All Rights Reserved
Commercial Use Strictly Prohibited

GROUNDED GRID CASCODE PHONO PREAMP

I get a lot of requests for a phono stage. Apparently, many audiophiles spin vinyl, or at least the ones that do are interested in my products. Most tell me that they have huge collections of LPs that they can't replace and thoroughly enjoy. Personally, I've been out of the vinyl spinning business for quite some time. To round out the book and provide an interesting project for those of you who enjoy your records, this one's for you. There are many variations and configurations that could evolve from this design. As of this writing, I have not decided exactly what I will put into production as a Transcendent Sound phono stage. My turntable and cartridge are definitely not "state of the art.". In fact, they aren't "state of anything." Before I can choose a final design, I am going to have to upgrade my front end. So in contrast to the other projects in the book, this one is still under development.

If you haven't guessed by now, I like to design power amps most of all. This is just my personal preference. All of us tend to excel at tasks that we enjoy the most. Vacuum tubes are best suited for power amp applications. The signal levels are high enough that tube noise and microphonics pose little if any threat to performance. As signal levels drop, tube noise begins to become a problem. At line level amplitudes, tube noise is still not much of a factor as is demonstrated by the quietness of my grounded grid line stage. When signal levels get down to a few millivolts, however, tube noise becomes a serious problem. There are ways to bring the noise down somewhat, but things can quickly get unwieldy. A dead end will be eventually reached where no further improvement is possible. The designer is faced with the task of creating an expensive, complicated component or just use a three-dollar integrated circuit and be done with it. It's a lot easier and less expensive to use some of the new low noise integrated circuits. In fact, a high performance, IC based phono stage suitable for moving coil applications could be completely assembled by the hobbyist for less than $75. I firmly believe that the most suitable technology should be used for any given application. It may very well be that a tube-IC hybrid design gives the best overall performance. I am publishing this design because it does work well and is an interesting configuration. I'm not ready to boast that this is the best tube phono stage ever designed.

In keeping with the theme of this book, all of the designs presented–including this one–are substantially different from the standard offerings of the industry. The sonic characteristics of the grounded grid preamp are so good, I theorized that perhaps the configuration would make a good phono stage.

The first stage of the circuit is a grounded grid driving a cascoded second stage. The signal must pass through the first cathode follower before any voltage gain occurs. This should theoretically add about 3 dB of noise. It may reduce the signal to noise ratio slightly but I was hoping for an improvement in clarity and resolution as a more desirable benefit. So there is some method to my madness. The unweighted signal to noise ratio measured about

60 dB. Using an "A" weighting filter should improve that figure to between 65 and 70 dB. The prototype measured was a raw, open chassis with no shielding or optimization of component placement or wiring. I didn't even use shielded wires for the signal paths. Building the preamp on an optimized ground plane circuit board with the proper shielding should definitely improve noise performance. So the noise figure, while not being equal to modern solid-state designs, certainly rivals any conventional tube design.

The circuit is basically comprised of two sections. The first section contains a cascode gain stage which is driven in a grounded grid fashion. The signal first travels through V1A and then drives the cathode of V1B. A cascode is the quietest gain stage that can be constructed with tubes. The first section provides an open loop gain of nearly 40 dB. This gain is partially used to create the RIAA equalization needed for correct playback. The equalization network is contained in a negative feedback loop that connects the output from the first stage to the grid of V2B. Components C6, C7, R11, R12 alter the voltage gain to match the RIAA requirements. The feedback loop consumes about 20 dB of forward voltage gain leaving a net mid-band stage gain of 20 dB. The nature of phono equalization requires 20 dB of bass boost at 20 Hz and 20 dB of treble cut at 20 kHz. That's why turntable rumble can be such a problem; any rumble will be amplified by 60 to 70 dB before it even gets to the power amp.

The second stage is just a simple triode amplifier with an active plate load as was used in the grounded grid line stage. It is not involved with shaping the response characteristics. No negative feedback is used in the second stage and neither is any global feedback. An additional phase inversion is needed to allow for global feedback and that would require a third gain stage, which is unnecessary. The second stage provides a gain of about 26 dB. The total gain through the preamp is then 46 dB. This is perfectly suitable for moving magnet and high output moving coil cartridges. If your cartridge outputs less than 2 millivolts nominal, then a step-up transformer or low-noise "head amp" will be necessary.

The amplifying tubes are the 6922. I used them because they have the lowest noise figure of any conventional vacuum tube. They are extremely popular with contemporary manufacturers. As you can see, I have not used them in any of my other projects. This is because I feel they impart a coloration to the sound that I find undesirable. I'm not interested in following industry fads.

The preamp, as originally constructed, is highly susceptible to acoustic feedback. Tapping on the chassis with the unit in front of a speaker can cause howling. This is not a good thing. The 6922 is supposed to be one of the tubes most resistant to microphonic effects. The acoustic feedback was accentuated by having the tubes installed on chassis mount tube sockets. If they were mounted on a printed circuit board, they would be greatly decoupled from chassis vibration. I can certainly see how makers of damping systems can verify the need for their products based upon improvements with tube phono stages. The caveat here is that whatever helps a phono stage will probably have no appreciable benefit on a power amp as stated previously in the section on damping systems. It would be interesting to enclose a tube phono stage inside a heavily damped speaker enclosure thickly lined with sound absorbent material. That's the kind of isolation needed to thoroughly eliminate acoustic feedback. An exhaust fan will be necessary to ventilate the enclosure. If you have to go to this much trouble, those integrated circuits start to look more attractive all the time.

The power supply provides three voltages. One is a regulated 12 volts for the filaments. The second is a regulated negative 200 volts for the input stage. The third voltage is a positive 235 volts for the plate supplies. It is critical that the supply voltages be clean and free from ripple and spurious noises. The advanced hobbyist can certainly investigate more sophisticated regulation schemes as well as individual stage regulation.

CHECKOUT

As with the grounded grid preamp, checkout primarily consists of making sure that the power supply voltages are present and the few circuit voltages are correct. There is nothing to adjust.

PARTS LIST

R1	47K
R2	33K-2W
R3	2K
R4	24K-2W
R5	47K
R6	68K
R7	20K
R8, R9	2K
R10	1K
R11	7.87K
R12	97.6K
R13	10K
R14	3K
R15	220K
R16	750
R17	5.1K-2W
R18	1.8K-2W
R19	1.5K-2W
R20	2K-2W
C1-C4	22MFD-350V
C5	.1MFD-400V, QUALITY FILM TYPE
C6	.03MFD-100V, 2%
C7	.01MFD-100V, 2%
C8	1MFD-400V, QUALITY FILM TYPE
C9-C16	22MFD-350V
C17	.01MFD-630V
C18	2200MFD-25V
D1-D4	1A-1000V DIODE
D5-D8	2A-200V DIODE
Z1, Z2	100V-5W ZENER
IC1	12V-1.5A GENERIC REGULATOR MUST HEAT SINK
T1, T2	120V-50MA TRANSFORMER
T3	12V-1A TRANSFORMER
S1	6A SWITCH
F1	1/2A SLOW BLOW FUSE
V1, V2	6922 VACUUM TUBE
V3	12AU7A VACUUM TUBE
V4, V5	6922 VACUUM TUBE

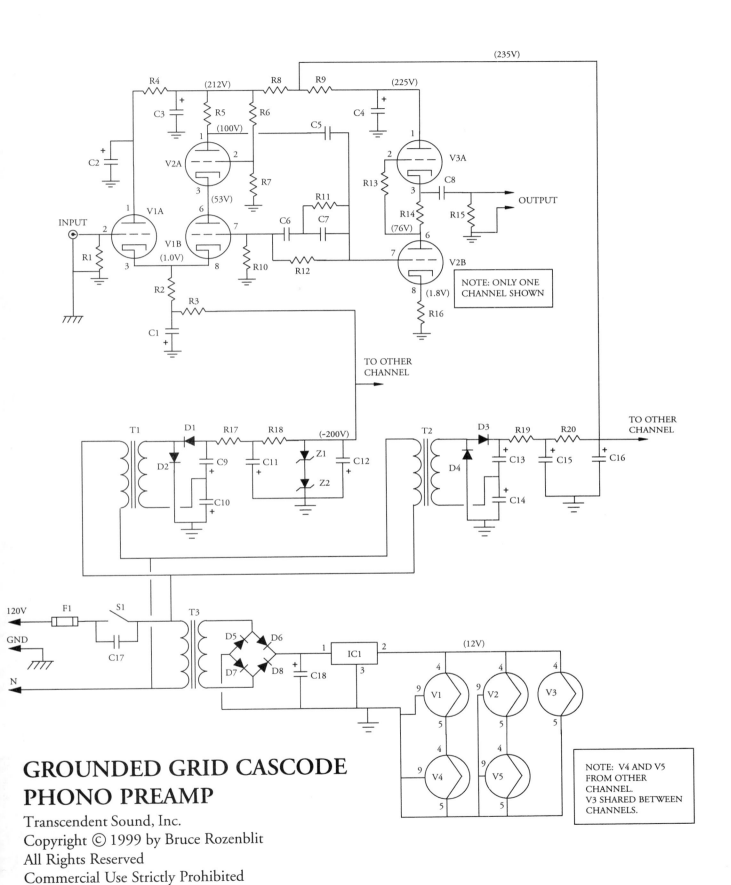

GROUNDED GRID CASCODE PHONO PREAMP

Transcendent Sound, Inc.
Copyright © 1999 by Bruce Rozenblit
All Rights Reserved
Commercial Use Strictly Prohibited

MY OTL PATENT

A patent award is one of the greatest acomplishments any design engineer can hope to achieve. Here's mine.

TRANSFORMERLESS OUTPUT VACUUM TUBE AUDIO AMPLIFIER
Inventor: Bruce M. Rozenblit
Patent No. 5,604,461

ABSTRACT

An audio amplifier including a series connected vacuum tube output stage for driving an audio loudspeaker. The preferred amplifier uses the output signal as a floating reference for an intermediate stage to prevent output stage degeneration for producing balanced push-pull drive signals, thereby eliminating the need for an output transformer or AC coupling.

Background-Field of the Invention

The present invention relates to the field of audio electronics and more particularly, to an audio power amplifier using a vacuum tube, series connected push-pull output stage for amplifying audio signals to drive a high fidelity loudspeaker.

Background-Description of Prior Art

In the field of audio electronics, the primary objective is to achieve faithful and accurate sound reproduction. Those individuals skilled in the art appreciate that vacuum tube-based audio equipment has sonic characteristics that are greatly desired by many audiophiles. One limitation of the traditional vacuum tube amplifier has been the necessity of an output transformer to provide impedance matching between the high impedance of the output stage and the low impedance of the loudspeaker. The output transformer is expensive to produce, and introduces a source of distortion and coloration into the sound. Specifically, the output transformer reduces the three dimensional imaging of the sound field-a characteristic highly sought after by audiophiles.

Prior art attempts to produce practical transformerless output vacuum tube amplifiers have been generally unsuccessful because of the following limitations:

(a) Extreme heat output–Waste heat output is more than double that of conventional transformer coupled amplifiers. Heat outputs for a typical 100 watt transformerless output amplifier range from 600 to 800 watts per channel. This is intolerable in a residential environment.

(b) Poor reliability–The output tubes are operated with high bias currents which greatly shortens tube life and stresses the tubes. High power transients push the tubes well beyond their maximum ratings. As a result, failures occur frequently.

(c) Poor bass sound–These types of amplifiers have always had a very weak and feeble bass sound when driving dynamic loudspeakers. It has been a commonly accepted adage in the audio industry that transformerless output amplifiers cannot supply bass drive equal to conventional transformer coupled amplifiers.

(d) DC instability–The DC level of the output stage is unstable forcing prior art designs to use an AC coupled output. The output capacitor increases cost, further reduces bass output because of its reactance and effective series resistance, and is audibly less desirable to audiophiles than a direct coupled connection to the loudspeaker.

The root cause of these problems has been the inability of prior art designs to provide truly balanced push-pull drive signals to the series connected output stage. The prior art techniques provide for equal amplitude push-pull drive signals but have unequal push-pull forward voltage gains. It is this gain unbalance that creates the conditions that cause these deficiencies.

Summary of the Invention

The present invention solves the prior art problems discussed above and provides a distinct advance in the state of the art. More particularly, the transformerless vacuum tube audio amplifier hereof provides a truly balanced drive with equal push-pull drive voltage gains through the system without the use of an output transformer or output coupling capacitor. Other preferred aspects are discussed further.

Objects and Advantages

Accordingly, several objects and advantages of the present invention are:

(a) Low heat output–The waste heat output of the amplifier is comparable to that of a conventional transformer coupled design. The heat output of the 110 watt configuration is in the range of 325 watts per channel which is about 100 watts greater than a conventional transformer coupled amplifier and half as much as prior art designs.

(b) Excellent bass sound–The apparent bass sound of the amplifier is virtually indistinguishable from a conventional transformer coupled unit.

(c) Solidly reliable–The low heat output of the design relieves much stress on the output tubes allowing them to last much longer and greatly reduces failure rates.

(d) Reduced maintenance–Since the output tubes last much longer, tube maintenance is reduced. Also, chassis temperatures are lower, allowing the internal components to last longer.

(e) Direct coupled output–The amplifier employs a truly direct coupled output to the loudspeaker which acoustically, is the most desired type.

Further objects and advantages of the invention will become apparent from a consideration of the drawings and ensuing description.

Brief Description of the Drawings
(Pages 119 and 120)

Figure 1 is an electrical schematic of the prior art illustrating the inability to achieve a truly balanced drive.

Figure 2 is a block diagram illustrating the preferred amplifier design according to the present invention.

Figure 3 is an electrical schematic of the amplifier of the preferred design.

Figure 4 is an electrical schematic of the preferred amplifier's power supply.

Detailed Description of the Preferred Embodiment

To illustrate the improvements of this invention, a detailed analysis of the prior art is necessary.

Figure 1 shows the typical prior art attempt to achieve a balanced drive. Bias circuitry has been omitted for clarity. Triodes VB and VC comprise the series connected output stage. Triode VA is a phase splitter used to generate two equal and out phase signals for driving the output stage. The gain of stage VA is essentially unity. The cathode of VC is at AC ground through the negative power supply.

The cathode of VB is not at ground potential. It floats on the output signal at node C. The drive signal for VB, SU, is in phase with the output signal SO. This causes degenerative feedback for the drive signal at node A. With no further compensation, the output stage is unbalanced.

The purpose of the Floating Regulator is to

provide compensation for the degenerative feedback at node A. The Floating Regulator shown is a three terminal series pass type. Since the regulator is referenced to the signal output at node C, the potential applied at node D is the summation of the regulator's DC output voltage (B+) and the audio output signal. The signal applied to node D is in phase with the drive signal at node A. This provides the exact amount of signal boost or positive feedback required to overcome the degeneration of VB.

The prior art attempt to provide a balanced drive seems successful but it is not. This circuitry does provide balanced drive signal amplitudes for the output stage but it has unequal forward voltage gains through the system.

The drive signal SL at node B is essentially equal and in phase with signal SI at node E. Due to the action of phase splitter VA, signal SL is constrained to equal SI. The compensation signal applied at node D is out of phase with SL. This causes negative feedback to be applied to node B. Forward voltage gain is then consumed at node B. The output stage is unbalanced because the forward voltage gain at node A has been increased by the action of positive feedback and the forward voltage gain at node B has been decreased by the action of negative feedback.

The unbalanced forward voltage gains through the system have many detrimental effects on performance, particularly regarding bass performance. When making steady state measurements into a load resistor, bass response seems satisfactory as -3db points at 10 Hz are easily obtainable. However, acoustic bass output is deficient.

The prior art technique as described above produces a DC offset in the output signal that is proportional to output signal amplitude. This pushes the loudspeaker cone in one direction which limits cone excursions reducing acoustic bass output. Bass drive requirements for reproducing music require large amplitude signals further exaggerating the problem.

The unequal forward voltage gains cause the amplifier's output impedance to differ between positive and negative going portions of the output signal. The damping factor is therefore unequal regarding compression and rarefaction strokes of the loudspeaker increasing the difficulty of the amplifier to control cone motion. These prior art deficiencies greatly reduce the acoustic bass output of the loudspeaker, which is easily apparent to those individuals skilled in the art.

A further complication caused by the unbalanced drive is increased distortion in the system. This has forced prior art designs to use large amounts of global negative feedback to reduce distortion and has also forced the use of high bias current operation for the output stage. Those individuals skilled in the art recognize that large amounts of global negative feedback decrease sonic quality particularly regarding imaging. Also, the large feedback consumes forward voltage gain thereby reducing the signal amplitude available to drive the output stage. This causes designers to opt for high bias current operation for the output stage because it requires less drive voltage. High bias current operation has the detrimental effects of causing DC instability requiring the use of an AC coupled output, greatly decreases tube life, and causes inefficiency that doubles the waste heat output over conventional transformer coupled designs.

All prior art attempts at achieving a balanced drive for the series connected output stage suffer from these performance limitations because they all apply the same compensation to both push-pull drive signals for the output stage causing unequal forward voltage gains through the system.

Figure 2 illustrates the preferred vacuum tube audio amplifier 10. Amplifier 10 includes gain stage 18, intermediate stage 20, output stage 22 and power supply 24. Gain stage 18 provides all of the forward voltage gain for the amplifier as well as generates two split phase signals for driving the output stage. Intermediate stage 20 applies positive feedback 26 from the output line 32 to only one of the split phase signals and none to the other. In this manner, drive signal degeneration for the output tube whose cathode is coupled to the output line is eliminated. No feedback compensation is applied to the drive signal for the output tube whose cathode is coupled to ground through the power supply.

The Amplifier

Figure 3 presents an electrical schematic of the amplifier 10 except for power supply 24 shown in Figure 4. Triode V1 (type 12AX7A) receives the audio signal at its grid. Resistor R1 (100K) is coupled between the grid of V1 and ground which provides an input impedance of 100K. Series-coupled resistors R2 (3K) and R3 (100 ohms) are coupled between the cathode of V1 and ground and create a negative bias voltage of 1.3 VDC by virtue of the idle current flowing through the tube. Capacitor C1 (100 mfd-25V), connected in parallel with resistor R2, provides an AC bypass to minimize local degenerative feedback in V1. Resistor R. (220K) is coupled between the +225 volt supply and plate of triode V1 where the output signal for V1 is developed. The voltage gain for V1 is about 70 times. The DC potential at the plate of V1 is about 125 volts which is used to set the bias of the next stage.

The node between resistors R2 and R3 provides the injection point for global negative feedback from output line 32 by way of resistors R9 and R10 (3K each) and capacitor C3 (1500 pf-500V). Switch S1 connects R10 in parallel with R9 and C3. With switch S1 closed, the amplifier has a forward voltage gain of 22 db. With S1 open, the amplifier has a forward voltage gain of 28 db. Capacitor C3 is sized to provide a slightly over-damped response to minimize ringing when driving highly capacitive loads such as electrostatic speakers.

Triodes V2A and V2B (type 12AU7A) are connected as a cathode coupled phase splitter. The cathodes of V2A and V2B are connected together and to one end of resistor R8 (15K-2W). The other end of R8 is grounded. The grid of V2A is directly coupled to the plate of V1. The grid of V2B is grounded through capacitor C2 (0.1 mfd-630V) and connected to the grid of V2A through resistor R5 (1M). Resistors R6 (33k-2W) and R7 (39K-2W) connect the plates of V2A and V2B to the power supply voltage of 480V. They are staggered in value to allow the phase splitter to produce two out of phase signals equal in amplitude. The phase splitter has a gain of about seven times and can produce highly linear outputs of 200 volts peak to peak. The DC potential at the plates of V2A and V2B is about 280 volts which is used to set the bias of the next stage.

Drive signal compensation stage 20 is comprised of two cathode followers, each direct coupled to one output of the phase splitter. Triode V3B has its cathode connected to ground through series coupled resistors R14 and R15 (both 33K-2W). The plate is connected to the 480V power supply. The grid is direct coupled to the plate of V2B. Capacitor C5 (1.0 mfd-630V) is coupled to the cathode of V3B and serves to isolate the DC voltage on the cathode from the negative bias voltage of the output stage. Triode V3A has its cathode connected to output line 32 through series coupled resistors R12 and R13 (both 33K-2W). The grid of V3A is direct coupled to the plate of V2A. The plate of V3A is connected to the output line 32 by way of a 450 volt shunt type voltage regulator comprised of three zener diodes Z1, Z2, and Z3 (all 150 volt-5W). The plate of V3A is also coupled to the 550 volt power supply through resistor R11 (10K-2W). Capacitor C4 (1.0mfd-630V) connected to the cathode of V3A serves to isolate the DC potential on the cathode from the negative bias voltage of the output stage. The voltage gain of stage 20 is about 0.95.

The output stage 22 is a series connected push-pull configuration. Triodes V4 and V5 (type 6C33C-B) yield a power output of 35 watts. An identical pair of triodes V6 and V7 with resistors R20 and R21 (both 100K) are added in parallel at the "X's" shown in figure 3 for the 110 watt design. The cathodes of V4 and V6 are connected to the ouput line. The plates of V5 and V7 are also connected to the output line. The grids of V4 and V6 are coupled to C4 through resistors R16 and R20 (100K each) respectively and receive the compensated drive signal. The grids of V5 and V7 are coupled to C5 through resistors R18 and R21 (100k each) respectively and receive the uncompensated drive signal. Negative bias voltage is injected into the node of C4, R16, and R20 through resistor R17 (100K) and into the node of C5, R18 (omitted from drawing), and R21 through R19 (100K). The bias voltage applied to R17 is -70 volts and is fixed. The bias voltage applied to R19 is nominally -240 volts and is adjustable. The bias voltage at R19 is 170 volts more negative than the bias at R17 to compensate for the cathodes of V5 and V7 being held at -170 volts. The loudspeaker is directly coupled to output line 32 without any intervening active or passive devices. Output stage 22 has a voltage gain of about 0.25. Resistor R32 (7.5-2W) and capacitor C23 (0.1mfd-630V) are series connected between the output line 32 and ground. They form a constant load impedance for ultrasonic frequencies.

Theory of Operation

Stage 18 is a conventional two stage voltage amplifier-phase splitter that has been used in vacuum tube audio power amplifiers since the 1940's. Critical performance parameters for stage 18 are that it must produce two linear out of phase audio signals, each with a maximum peak to peak amplitude of 200 volts, and have a forward voltage gain of about 500 times. Those individuals skilled in the art appreciate that a circuit with lower or higher voltage gain will still work but will alter the performance characteristics of the amplifier. It is also important that section 18 be direct coupled because any additional AC couplings would cause low frequency phase shift leading to instability and motorboating. Different vacuum tube circuit configurations are possible so long as they provide sufficient gain, amplitude, and phase linearity. It is recognized by those individuals skilled in the art that a circuit using solid state components could also be used for section 18. This would alter the sonic characteristics of the amplifier but is a viable way to drive the series connected vacuum tube output stage.

Stage 20 is the intermediate stage that contains the drive signal compensation which is the basis of the invention. Two cathode followers are required to provide symmetry and for equal phase shift and impedance for the two push pull drive signals. Triode V3A applies the required feedback compensation to triodes V4 and V6 while triode V3B provides no signal compensation to triodes V5 and V7. It is noted that cathode follower V3B could be omitted and the circuit will still function. If omitted, the time and phase characteristics of both push-pull drive signals will not be equal which will cause a deterioration in performance. Stage 20 must have a gain of unity because 100% positive feedback compensation is applied to overcome the 100 % degenerative feedback in the output stage. Cathode followers are used because they have a voltage gain of essentially unity. Vacuum tubes have been used for stage 20 but it is recognized by those individuals skilled in the art that solid state devices could also be employed.

It is recognized by those individuals skilled in the art that two closed loop unity gain stages consisting of multiple active devices could be used in place of the cathode followers. These closed loop stages could be constructed of vacuum tubes or solid state devices.

In order for a triode to operate as a unity gain cathode follower, its plate must be connected to ground through a very low impedance path. In a conventional application this path would be provided by the power supply. Since the cathode resistor of V3A is not grounded but connected to the output line, this low impedance path must be incorporated by some other means. One technique that has been tried is to connect the plate of V3A to the output line through a large capacitor. This produces excellent measurable results but the sound quality of the amplifier is very harsh and unpleasant. It has been discovered that the DC potential between the plate of V3A and the output line must remain fixed in order to produce high quality sound. A voltage regulator can provide both a low impedance path as well as fixing the DC potential with a high degree of accuracy.

The amplifier uses a two terminal shunt type regulator. A three terminal series pass type regulator will also provide excellent results but is more complicated and increases cost. It is recognized by those individuals skilled in the art that the voltage regulator could be comprised of zener diodes, gas tubes, vacuum tubes, or other solid state devices. The zener diodes used in the invention provide the lowest cost method to implement the voltage regulator with a very high degree of reliability.

The quiescent voltage drop across resistor R11 is a critical parameter. As the output line goes positive with the audio signal, the voltage on the cathode and plate of V3A rise in step with it. The voltage drop across R11 decreases by the same amount as the rise on the output line. Hence sufficient voltage headroom must be present in the drop across R11 or else premature clipping will occur. The amplifier uses a voltage drop of 100 volts across R11 which is sufficient to achieve full power and not drive the cathode follower into cut off.

The primary function of grid resistors R16, R18 (omitted from drawing), R20 and R21 is to prevent parasitic oscillation from occurring caused by the low impedance of the cathode followers and the input capacitance of the output tubes. A value of 10K ohms provides stable operation. It has been discovered that a 10K value causes the sound to be extremely harsh and irritating. The -3db point for the output stage with 10K ohm grid resistors is about 250 kHZ. When these resistors are increased to 36K ohms, the sound quality begins to markedly improve. The amplifier uses 100K ohm grid resistors which cause the sound to be very pleasant and musical. The -3db point for the output stage with 100 K ohm grid resistors is about 30 kHZ. This improvement in sound quality with large value grid resistors is unique and specific to the 6C33C-B type triode used in this amplifier. No other vacuum tubes have been found that duplicate this effect.

The Power Supply

Power supply 24 shown in figure 4 includes output section 34, drive section 36, filament heater section 38 and bias section 40. Transformer primaries are conventional (not shown). Output section includes a center tapped 230 volt winding from transformer T1 feeding a 35 amp-600V full-wave bridge rectifier B1. Series connected capacitors C6 and C7 (6000mfd-200V each) provide the filtering. Resistors 22 and 23 (20K-5W each) are discharge resistors for capacitors C6 and C7. Fuses F1 and F2 (4 amps fast blow each) protect the output stage and speaker against faults. Capacitors C8 and C9 (1 mfd-250V each) are bypass capacitors needed to create a good AC ground at high frequencies. The output voltages are +/- 170 volts at idle and +/-155 volts at full load.

Drive section receives 200 VAC from transformer T1. Diodes D1 and D2 (1A-1000V each) form a voltage doubler in combination with capacitors C10 and C11 (22 mfd-350V each). The resultant 550 volts DC is applied to V3A. Resistor R24 (5K-5W) reduces the voltage to +480 VDC for V2A, V2B, and V3B where capacitors C12 and C13 (210mfd-350V each) provide the filtering. Resistors R25 and R26 (300K each) equally divide the voltages across capacitors C12 and C13. Capacitor C14 (0.1mfd-630V) provides a high frequency bypass for V3A and capacitor C15 (0.47mfd-630V) provides high frequency bypass for V2A, V2B, and V3B. Resistor R27 (620K) drops the voltage to +225 VDC for V1 and capacitor C16 (22mfd-450V) provides the filtering while capacitor C17 (0.1mfd-630V) provides a high frequency bypass.

Filament heater section 38 receives 20 VAC from Transformer T1 which is supplied to full wave bridge B2 (35A-200V). Capacitor C18 (20,000mfd-40V) provides the filtering with the output being 25 VDC to the filaments of the vacuum tubes in amplifier 12. The tubes having 12.6V filaments are wired in series to utilize the 25 VDC power supply. Filament connections are conventional (not shown).

Bias section 40 provides the bias voltages for the output tubes at -240 VDC and -70 VDC. Stable bias is important and because of this, this power supply section is supplied with its own separate transformer T2 (120V-50mA) to provide better regulation.

In section 40, diodes D3 and D4 (1A-1000V) form a voltage doubler in combination with capacitors C19 and C20 (22mfd-350V each). Resistor R28 (39K-2W) drops the voltage to -240 VDC for the bias on tubes V4 and V7. Capacitor C21 (22mfd-350V) provides the filtering. Potentiometer R29 (10K-0.5W) provides a range of adjustment of 30 volts to balance output stage 22 for 0 volts on output line 32. Resistors R30 (53.6K) and R31 (24K) provide a fixed potential of -70 VDC for the bias on tubes V4 and V6. Capacitor C22 (22mfd-350V) provides the filtering.

Summary, Ramifications, and Scope

Accordingly, it can be seen that the amplifier of this invention provides truly balanced drive signals to the series connected vacuum tube output stage resulting in a low heat, reliable, transformerless output vacuum tube amplifier with excellent bass sound.

Although the description above contains many specificities, these should not be construed as limiting the scope of the invention but as merely providing illustrations of some of the presently preferred embodiments of the invention. For example, any of the circuit stages could be constructed of solid state devices or use vacuum tube based circuits of differing topology that achieve the same functional results; the floating voltage regulator could be three terminal series pass as well as two terminal shunt type or replaced with a capacitor; driver circuits of lessor or greater voltage gain could be used; the output stage could use a single-ended power supply with an AC coupled output; different output tubes could be used; high bias current or class A type output stage could be used; a transformer could be connected to the output stage for further output impedance reduction; closed loop feedback could be eliminated; different power supply voltages could be used.

Thus the scope of the invention should be determined by the appended claims and their legal equivalents, rather than by the examples given.

CLAIMS:

1. An audio amplifier comprising:
a gain stage including means for receiving an audio input from source thereof and responsive to said input for producing first and second amplified audio signals 180 degrees out of phase from one another;

an intermediate stage including first and second intermediate unity gain sections having means for respectively receiving said first and second amplified audio signals and responsive thereto for producing respective first and second intermediate signals;

an output stage including first and second series connected vacuum tubes coupled in a push-pull configuration, and including means for setting the quiescent current of said first and second series connected vacuum tubes, and for receiving and responding to said first and second intermediate signals for producing an output signal on an output line, said output stage being operable for driving an audio loudspeaker coupled with said output line without a transformer and without a capacitor between said output stage and said audio loudspeaker;

said first and second series connected vacuum tubes being coupled for respectively receiving at the grids thereof said first and second intermediate signals and responsive thereto for respectively producing said output signal;

the cathode of said first series connected output tube being coupled with said output line so that said cathode follows said output signal resulting in degeneration of said first intermediate signal at said grid of said first series connected vacuum tube relative to said output line;

and means for compensating for said degeneration including means for providing said output signal as a voltage reference for said first intermediate unity gain section so that said output signal is superimposed only on said first intermediate signal thereby compensating for said degeneration.

2. The amplifier as set forth in claim 1, said gain stage including means for providing a gain of about 500 for said first and second amplified audio signals relative to said input.

3. The amplifier as set forth in claim 2, said gain means including vacuum tubes.

4. The amplifier as set forth in claim 1, said means for said first and second intermediate unity gain sections including vacuum tube cathode followers.

5. The amplifier as set forth in claim 4, the plate of said first intermediate vacuum tube cathode follower be coupled to said output line by means of low impedance device.

6. The amplifier as set forth in claim 5, said low impedance device further providing a constant voltage drop.

7. The amplifier as set forth in claim 5, said low impedance device means including a

voltage regulator.

8. The amplifier as set forth in claim 6, said constant voltage drop of about 450 volts direct current.

9. The amplifier as set forth in claim 7, said voltage regulator including at least one shunt type zener diode.

10. The amplifier as set forth in claim 4, the cathode of said first intermediate vacuum tube cathode follower be coupled to said output line by way of resistance so that said output line provides a floating voltage reference to said cathode of said first intermediate vacuum tube cathode follower.

11. The amplifier as set forth in claim 4, the cathode of said second intermediate vacuum tube cathode follower be coupled to ground by way of resistance.

12. The amplifier as set forth in claim 1, said first and second intermediate signals each presenting about 200 volts peak to peak.

13. The amplifier as set forth in claim 1, said first and second series connected vacuum tubes respectively receiving said first and second intermediate signals by way of respective first and second grid resistance, each being greater than about 36K ohms.

14. The amplifier as set forth in claim 13, said first and second grid resistance each being about 100K ohms.

15. The amplifier as set forth in claim 13, said first and second series connected vacuum tubes being type 6C33CB triodes.

16. The amplifier as set forth in claim 1, further including feedback means for providing negative feedback from said output line to said gain stage.

17. The amplifier as set forth in claim 1, said output stage including means for operating in substantially class B mode.

18. The amplifier as set forth in claim 1, further including at least one pair of said series connected vacuum tubes.

19. The amplifier as set forth in claim 1, said means for setting said quiescent current of said first and second series connected vacuum tubes includes injecting negative bias voltage into said grids of said first and second series connected vacuum tubes, respectively.

20. The amplifier as set forth in claim 1, said output stage further includes means of setting quiescent voltage on said output line to zero volts by fixing said negative bias voltage on said grid of one said series connected vacuum tube and varying said negative bias voltage on said grid of other said series connected vacuum tube.

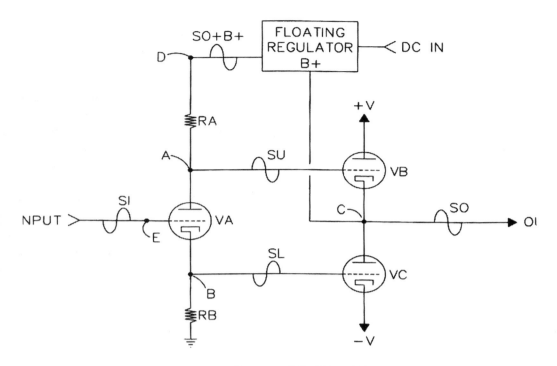

FIGURE 1 — PRIOR ART

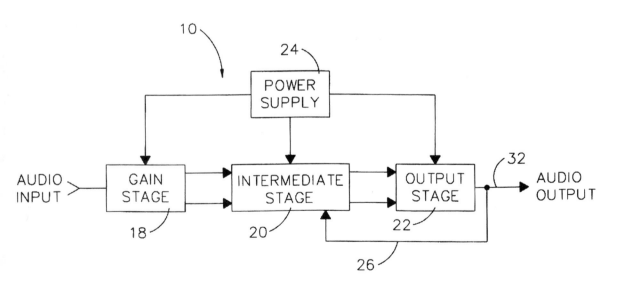

FIGURE 2

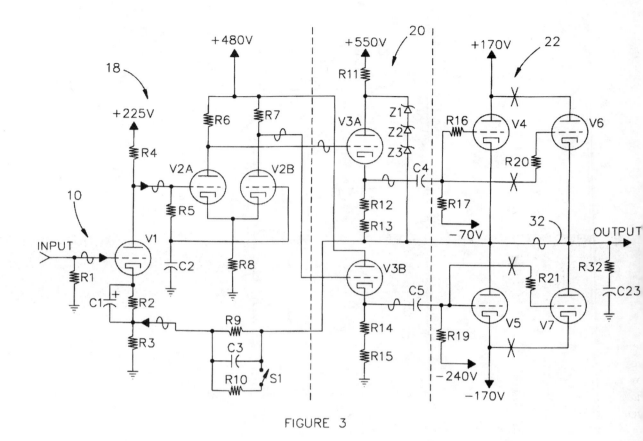

FIGURE 3

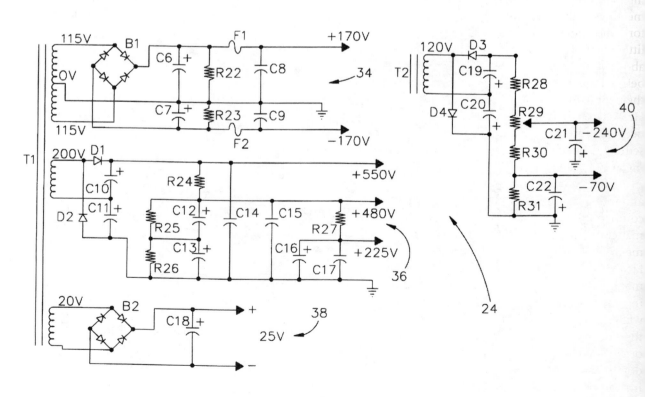

FIGURE 4

CONCLUSION

Well, what do you think? Did I make any sense? Or are you waiting for a *real* expert to tell you how to pass judgement on me? Hopefully, by acquiring knowledge and understanding, *you have become the expert.* You just don't realize it yet.

At one of my earliest jobs in the engineering field, my boss told me that when I go to meetings with other engineers, vendors, contractors, and clients, I will be surprised as to how little most of these "authorities" really know about their respective fields. He told me that before long I would know more than most of the people that I would be doing business with. He was exactly right. I soon concluded that becoming an authority has as much to do with selling yourself as it does with actually possessing any special knowledge.

No other manufacturer has such a diverse and innovative product line. Look at what the competition has to offer and compare performance, features, and price. (You've got to expect me to hype my products a little!) Other manufacturers will show you numerous endorsements, heavy sheet metal boxes, lots of advertising, exotic parts, and big price tags. Ask them what's in the pretty box. Ask them to show you their design. Why is it any better than any others? What features in use and performance does it have that would cause you buy it? Instead what you will hear is that *somebody* says it's the best in the world. Why does that "somebody" know any more than you do?

I have shown the actual schematics with circuit explanations of all of my products. What else could I possibly do to prove that all my equipment is completely different from every other manufacturer? A schematic is like the DNA of electronics. It details exactly how a product has been designed. The schematic is not an expression of an opinion about something. It is a statement of fact describing how an electronic component is put together. I have laid my products bare for all to see and evaluate.

Perhaps I've got it all wrong. Maybe the buying public couldn't care less about the design and construction of their audio equipment. My life's work could very well be an ongoing exercise in futility in that I'm trying to sell innovation and superior performance in affordable equipment. The competition just might have the right idea after all. Hire an industrial designer to come up with some far out looking enclosure, use the most expensive, exotic parts available to create an aura of exclusivity, use a common circuit with reasonably good performance, pump the price way up to create the illusion of status of ownership and sales will certainly skyrocket. This must work because that is exactly how 90+ percent of high-end audio gear is marketed. The ads virtually never say how or why their products are different. I remember one amplifier ad that just showed off the amp's binding posts. Another featured a photo of the power switch. As ridiculous as these marketing techniques appear to me, manufacturers certainly wouldn't use them if they didn't generate sales.

The very nature of audio creates an environment that is perfectly suited for this type of marketing. If a toaster doesn't make toast, anyone can see that. If audio equipment has fabulously mediocre performance, who's going to fault it? It still sounds fine. There's nothing demonstrably wrong with it. Nice sounding music pours out of the speakers. No one receives an injury or disease from using the system. No crime has been committed. However, there is a great deal of equipment available (usually made by small manufacturers) that sounds a lot better for one half to one third the cost. So what? People often buy perfume that costs 50 cents to make for $50 a bottle. I myself have occasionally purchased chic bottled water from France, and I'm so tight I squeak when I walk. Hey, it was a treat. It's my money and if I want to waste some I will! That's my point exactly. It's your money and you have every right and privilege to spend it any way you want. The question I have for you is: Do you *really* know what it is you are buying? If the answer to that question is a concern, then I sincerely hope this book has helped you.

Once the endorsements are in place and the mass marketers get a hold of you, money flows from your wallet to theirs. All the masters of selling need from the circuit designer is a decent level of performance. They can handle the rest by creating the necessary images and illusions that get the consumer to write the check. Lots of checks are written and lots of expensive, average-sounding equipment gets sold. The buyer doesn't know the difference and is happy with his new status symbol. As long as the system plays loud and clean with lots of bass, it really doesn't matter if the sound is just passable. *Audio has therefore become a business that primarily sells illusions.*

There are absolutely no mystical techniques, procedures or materials that can improve the performance of stereo equipment. It's all science and engineering. The design is what makes the equipment perform as it does. Why is it that people expect their doctor to use science to heal them, but expect their audio equipment to employ mystical devices to sound better? That's another question you will have to answer for yourself.

There seems to be a continuous battle in the media between engineering types and nontechnical types concerning why, or if, some component changes alter the character of the sound. The tech types usually take the position that if a change in the signal can't be measured, then any perceived differences are purely imagined by the listener. The nontechnical types (the overwhelming majority) often take the position that this scientific stuff has nothing to do with it because there is "something else" going on. The following argument summarizes my views on the subject.

A loudspeaker is an electric motor. The sound that emanates from it is totally dependant upon the electric signal fed into to it. That's a fact. The electrical signal makes the speaker cone move and cone movement is what generates the sound. Any equipment alterations ahead of the speaker that result in a change in the sound *absolutely* must alter the electric signal in some manner. Otherwise, no change in the sound could possibly occur. All sonic characteristics are subordinate to the electrical signal. These signals are physical entities and their nature and behavior *absolutely* must obey the laws of physics. All physical phenomena behave according these laws. They are the laws of nature.

All electrical signals possess certain parameters that totally define them. These parameters are:
1. *The spectral content of the signal.*
   These are all of the discrete frequencies of which the signal is comprised.
2. *The amplitude of the spectral components.*
   These are the sizes or strengths of all of the individual discrete frequencies.
3. *The time distribution of the spectral components.*
   These are the times when the individual discrete frequencies occur.

There are no other factors. This is not my opinion but a proven mathematical fact and verifiable with physical experiments. Understanding the totality of these three parameters reveals complete linkage between science and sound reproduction.

Therefore:

Any device that is purported to change the nature or character of the sound, must in some way alter one or more of these parameters. No other possibilities exist.

Therefore:

Any device that claims to alter the nature or character of the sound but does not possess any mechanism to alter any of these parameters can have no possible effect.

I have concluded that the measurement of subtle changes in sonic character can be very difficult because of the dynamic nature of sound. The measurement problem is compounded because, even if you can determine what to look for and have the equipment and technology to find it, you are looking for a moving target that won't sit still. Static measurements, therefore, don't tell the whole story. This is a common pitfall that influences many tech types. They can't measure a difference so they conclude there is no difference. The problem is that they are looking for the wrong things. That's why they can't find them.

The nontechnical majority also falls victim to its own set of predetermined notions. I have had many conversations with customers who have told me about major improvements to the sound when a binding post was changed, a cable direction was reversed, or a short length of wire was replaced. Some of these very same people have operated my OTLs for days, or even weeks, with a blown output tube, a blown output stage fuse, or even with the driver tubes in the wrong sockets and *did not* realize there was a problem. Eventually they noticed a reduction in maximum power output. That's when I'd get the call. Then I'd get to tell them that my amps are so wonderful that they sound great even when they aren't operating properly.

How can they listen to equipment that is grossly malfunctioning and not notice the problem while at the same time be convinced that questionable enhancements are so productive? I can only conclude that many of these "benefits" result from a self-fulfilling prophesy. The listener wants to experience an improvement, so an improvement occurs.

For a passive device to alter an electrical signal it must change the resistance, capacitance or inductance acting on the signal. No other characteristics exist. The consumer should always investigate advertised claims with regard these three characteristics. If no mechanism can be identified that alters these characteristics, then the product's performance is highly suspect.

In the spring of 1996, I participated in a trade show in New York City. This was essentially my debut to the audiophile community. It was a memorable event for me and I'd like to share with you one of its most unforgettable moments. I noticed that virtually all of the African-American audiophiles that came into my room stayed for at least one half hour. Many stayed longer. The opposite was true of the caucasion audiophiles. I was lucky if they would stay for five minutes. The differences between reactions in regard to listening time was striking when broken down by race.

On the last day of the show, I was engaged in a lengthy conversation with an African-American man about stereo systems. I felt that we had connected personally, man to man, so I asked him, "Why is it that most of the white people that come into my room can't sit still for five minutes and almost all of the black folks stay for more than half an hour and many even stayed for a whole hour?"

He didn't have to scratch his head and think about it. His answer came freely, quickly, without any hesitation or reservation. His answer, which I highly paraphrase and condense, went something like this: "We have to think for ourselves. We don't care what anyone else says or what the brand name is. We're just interested in the performance."

With all of the other big name systems in the building, these audiophiles stayed with me, a company they had never heard of, for an extended period because they enjoyed the quality of the musical experience so much. The music was more important to them than anything else.

Tube equipment is much easier to put together than solid-state. The circuitry is inherently simpler. All of the parts are much larger and easier to manipulate. The very nature of tube circuitry facilitates easy point-to-point wiring. Circuit boards are completely unnecessary for single unit construction. Tube circuits are extremely rugged and can withstand a tremendous amount of abuse. Short out a solid-state circuit with a test probe and all the semiconductors can easily be ruined. I've had that happen more than once. It's very embarrassing to go to the parts store to buy two replacement transistors and come back that afternoon and buy them again plus six more.

No matter what your opinions of tube equipment are, this field is by far the easiest way to get involved with building your own stereo equipment. The costs can be extremely low and the performance is very high. I especially hope the younger readers will get involved and build up a system and even a pair of speakers. I've always thought that making inexpensive useful things that worked well was a lot of fun.

So where do we go from here? Has Rozenblit designed his last circuit? No way! I'm just getting warmed up. I design because I enjoy doing it. Actually, the process of design is more important to me than what I'm designing. The project material drained a lot of creative energy from me and I need some time to recharge. It's no easy task to develop four or five radically new designs in a short period of time. As soon as I get good and bored, I'll think of something new to come up with. This is too much fun, particularly when the established authorities are convinced that nothing new can possibly be developed. We just showed them how wrong they are.

REFERENCES

1. *Fundamentals of Physics*, David Halliday and Robert Resnick, John Wiley and Sons, 1974.

2. *Electric Networks*, Hugh Skilling, John Wiley and Sons, 1974.

3. *Standard Handbook for Electrical Engineers*, Donald Fink and Wayne Beaty, McGraw-Hill, Eleventh Edition, 1978

4. *Electronics Engineers Handbook*, Donald Fink and Donald Christiansen, McGraw-Hill, Second Ed., 1982.

5. *Linear Circuits*, Ronald Scott, Addison-Wesley, 1960.

6. *Engineering Electro-Magnetics*, William Hayt, Jr., McGraw-Hill, Third Edition, 1974.

7. *Practical Transformer Design Handbook*, Eric Lowdon, TAB Professional Books, Second Edition, 1989.

8. *Sound System Engineering*, Don and Carolyn Davis, Howard Sams & Co., Second Edition, 1987.

9. *Reference Data For Radio Engineers*, Howard Sams & Co., Sixth Edition, 1975.

10. *Handbook for Sound Engineers*, Gene Ballou, Editor, Howard Sams & Co., 1987.

11. *Jensen Transformer Catalog*, Application Notes 002, 003, and 004, Bill Whitlock, 1995.

12. *Beginner's Guide To Tube Audio Design*, Bruce Rozenblit, Audio Amateur Press, 1997.

13. *From Compass to Computer*, W. A. Atherton, San Francisco Press, 1984.

14. *Active Networks*, Vincent C. Rideout, Prentice-Hall, Inc. 1954.

15. *Principles of Electron Tubes*, Herbert J. Reich, Audio Amateur Press, 1995.

16. *The Radiotron Designers Handbook*, F. Langford-Smith.

PARTS SOURCES

The following companies can supply you with all of the parts needed to build these projects, including tools and test equipment. They are in business to service the hobbyist. Their catalogs are free and most are published on the web.

Antique Electronic Supply
602-820-5411
Fax: 800-706-6789
www.tubesandmore.com

Hammond Manufacturing
716-631-5700
Fax: 716-631-1156
www.hammondmfg.com

Parts Express
800-338-0531
Fax: 513-743-1677
www.parts-express.com

Old Colony Sound Lab
603-924-6371
Fax: 603-924-9467

Mouser Electronics
800-992-9943
Fax: 817-483-6899
www.mouser.com

New Sensor Corp.
800-633-5477
Fax: 212-529-0486
www.sovtek.com

Digi-Key
800-344-4539
Fax: 218-681-3380
www.digikey.com

MCM Electronics
800-543-4330
Fax: 800-765-6960
www.mcmelectronics.com

Tech America
800-877-0072
Fax: 800-813-0087

Avel Transformers
860-355-4711
Fax: 860-354-8597

Radio Shack

WANT TO BUY THE REAL THING?

*Now that you've read about it... Understand and appreciate it...
Want it more than anything... Go ahead–Buy it!*

The Transcendent OTL
Monoblocks–$4495 pr.
Stereo–$2295 each.

Class "A" Rated–Stereophile 1998
Golden Ear Award–
The Absolute Sound, Dec. 1998

The Grounded Grid Preamp
Only $899.

The Super Compact 150
Only $1199 each monoblock.

Visa/MasterCard Accepted–No Risk Two Week Trial

Factory Direct Pricing–
Your Best Value in High-End Audio

For the latest developments–
Check out the web page at:
www.transcendentsound.com

Phone: 816-333-7358
Fax: 816-822-2318

Transcendent Sound, Inc.
P.O. Box 22547
Kansas City, MO 64113